International Virtual Conference on Recent Trends in Materials Science (ICRTMS 2021)

IOP Conference Series: Materials Science and Engineering
Volume 1233

Online
22 October 2021

Editors:

Augustine George
A. Sevuga Pandian
M. Ambrose Rajkumar
Ajith S. Kumar

ISBN: 978-1-7138-5436-4
ISSN: 1757-8981

Printed from e-media with permission by:

Curran Associates, Inc.
57 Morehouse Lane
Red Hook, NY 12571

Some format issues inherent in the e-media version may also appear in this print version.

Institute of Physics
Dirac House, Temple Back
Bristol BS1 6BE UK

Phone: 44 1 17 929 7481
Fax: 44 1 17 920 0979

techtracking@iop.org

Additional copies of this publication are available from:

Curran Associates, Inc.
57 Morehouse Lane
Red Hook, NY 12571 USA
Phone: 845-758-0400
Fax: 845-758-2633
Email: curran@proceedings.com
Web: www.proceedings.com

TABLE OF CONTENTS

IOP Conf. Series: Materials Science and Engineering 1233 (2022) 011001 doi:10.1088/1757-899X/1233/1/011001

Proceedings of the International Virtual Conference on Recent Trends in Materials Science ICRTMS 2021

October 22nd, 2021

Organized by

DEPARTMENT OF PHYSICS
KRISTU JAYANTI COLLEGE (AUTONOMOUS)
K. Narayanapura, Bangalore – 560077
Karnataka, India

Editors

Rev. Dr. Augustine George
Prof. A. Sevuga Pandian
Dr. M. Ambrose Rajkumar
Dr. Ajith S Kumar

IOP Conf. Series: Materials Science and Engineering 1233 (2022) 011001 doi:10.1088/1757-899X/1233/1/011001

Message From Principal

The role of materials science in the process of development is very important and crucial. It plays a vital role in this cutting edge technological era. Materials science is leading from the front to reach the excellence in every walk of life. An innovative invention and discoveries in the field of materials science changes the momentum of all fields including economics and finance marketing of our globe. The discipline of Physics of materials is one such technology with ancient roots. The fundamental feature of human intelligence is creativity. New ideas can be created by producing novel combinations of familiar ideas, by exploring the potential of conceptual spaces and by making transformations that enable the generation of previously impossible ideas.

I congratulate the Department of Physics for the initiatives taken to bring the researchers across the nation in a common platform. I am confident that this conference will provide a stage to present and share the research findings for further enhancement and perfection. I am delighted to note that the accepted papers are being published in *IOP Conference Series: Materials Science and Engineering (MSE) – a* SCOPUS indexed journal.

I wholeheartedly appreciate all the sincere efforts of the entire team of ICRTMS 2021 and wish them a grand success.

IOP Conf. Series: Materials Science and Engineering 1233 (2022) 011001 doi:10.1088/1757-899X/1233/1/011001

Message from Dean

Materials Science brings a lot to this universe ranging from biosensor to automatic lander in space vehicles. It contributes in all area of science, such as physics and engineering to make better, more useful, and more economical and efficient "stuff". We could realize the impact of new materials in science and Engineering, because of the interdisciplinary nature of materials science.

ICRTMS 2021 strives to be one of the most far reaching meetings concentrated on the different aspects of Physics and Materials Science. It is certain to be the stage for scholastic and industry experts to examine ongoing advancement in the fields of physics, materials science and allied fields. The reason for this gathering is to make a logical commitment to the field of physics through dialogues and publications.

Special appreciation to the convenors and the organizing committee and. We trust the delegates can acquire useful information from the proceedings. We look forward to the contributions during discussions and publications providing the foundation for future developments in this field.

On behalf of the Organizing and Advisory Committee we take great pleasure in welcoming the delegates to the conference to foster the progress in the field by contributing with your expertise.

IOP Conf. Series: Materials Science and Engineering 1233 (2022) 011001 doi:10.1088/1757-899X/1233/1/011001

Message from Conference Chair

Technology has ignited the minds of every human to explore the new methods which lead to the new innovations and discoveries in every field. The research in the field of science and engineering has done tremendous changes in human life. In future, we could expect research in materials science would help us to use non-conventional energy sources completely for generation of electricity and even other form of energy.

The International Virtual Conference on Recent Trends in Materials Science (ICRTMS 2021) conceives to unite interdisciplinary analysts researching the interface between Physics and Materials Science. Discussions as this will be instrumental in creating novel strategies for research and development. We sincerely believe this virtual gathering will be a platform for scientific exchange of ideas and to provide an opportunity for the dissemination of the latest results in research among the range of experts in the field Physics and Materials Science.

I sincerely hope that one day of healthy interactions and proactive exchange of ideas will prove to be fruitful and contribute immensely to our mutual growth. I congratulate the participants, Delegates, Organizing Committee, faculty members and students for their outstanding contribution in organizing this conference successfully.

IOP Conf. Series: Materials Science and Engineering 1233 (2022) 011001 doi:10.1088/1757-899X/1233/1/011001

Message From Conveners

Materials science is an interdisciplinary field which incorporates physics, chemistry, and engineering. It covers the design and fabrication of new materials for the development of modern devices and applications in a variety of research areas including nanoscience, biotechnology, energy storage and other functional materials.

This virtual conference aims at disseminating knowledge in the domain of Materials Science. This Conference will be an opportunity to reflect on what has been happening in the field of Science and Engineering, to share the experiences, perspectives and responses to the challenges by the scientific community, and at the same time to give thought to setting new directions for the future. The Conference aims at bringing together Physicists, Research Scientists and Research Scholars to discuss their latest findings in the areas of the physics and chemistry of advanced materials in Medical, Engineering, Nanotechnology, Environmental Technologies, Biomaterials and Green Chemistry. It is also a forum for the dissemination of original research findings, new ideas and practices.

We extremely thank our management, Keynote speakers, advisory committee members, participants, reviewers, session chairs, organizing committee Members and all those who have helped us to organize this International Conference and to make it a success.

We thank Indian Association for Crystal Growth (IACG) and our publication partner IOP Science Publishing for publishing the accepted papers in the *IOP Conference Series: Materials Science and Engineering (MSE) – a* SCOPUS indexed journal.

Dr. Ambrose Rajkumar M.

Dr. Ajith S kumar.

IOP Conf. Series: Materials Science and Engineering 1233 (2022) 011001 doi:10.1088/1757-899X/1233/1/011001

Conference Committee

CHIEF PATRON

Fr. Dr. Augustine George, CMI

Principal

ADVISORY COUNCIL

Fr. Emmanuel P. J., CMI

Director, Kristu Jayanti College of Law

Fr. Joshy Mathew, CMI

Fr. Deepu Joy Parayil, CMI

Dr. Calistus Jude AL,

Dean, Faculty of Sciences

CONFERENCE CHAIR

Fr. Lijo P Thomas

Financial Administrator

CONFERENCE CO-CHAIR

Prof. Sevuga Pandian. A

Head, Dept. of Computer Science (UG)

CONFERENCE CONVENERS

Dr. M. Ambrose Rajkumar

Asst. Professor, Dept. of Physics

Dr. Ajith S Kumar

Asst. Professor, Dept. of Physics

Dr. Mayank Pandey

Asst. Professor, Dept. of Electronics

ADVISORY COMMITTEE

Prof. P. Ramasamy

President, Indian Association for Crystal Growth, SSN College of Engineering, Kalavakkam, Tamil Nadu.

Dr. S. Angappane

Scientist, Centre for Nano and Soft Matter Sciences, Bangalore.

Dr. D. Prem Anand

HOD of Physics, St. Xavier's College (Autonomous), Palayamkottai, Tamil Nadu.

Dr. R. Ramesh Babu

Associate Professor, Department of Physics, Bharathidasan University, Tiruchirappalli, Tamil Nadu.

Dr. M. Senthil Pandian

Research Scientist, SSN College of Engineering, Kalavakkam, Tamil Nadu.

Dr. G. Markandeyulu

Professor, Indian Institute of Technology Madras, Chennai.

Dr. V. Srinivas

Professor, Indian Institute of Technology Madras, Chennai.

Dr. P. S. Anil Kumar

Professor, Indian Institute of Science, Bangalore.

Dr. Satyam Suwas

Professor, Indian Institute of Science, Bangalore.

Dr. S. Anbarasu

Asst. Professor of Physics, St. Joseph's College (Autonomous), Tiruchirappalli, Tamil Nadu..

IOP Conf. Series: Materials Science and Engineering 1233 (2022) 011002 doi:10.1088/1757-899X/1233/1/011002

Peer Review Statement

All papers published in this volume have been reviewed through processes administered by the Editors. Reviews were conducted by expert referees to the professional and scientific standards expected of a proceedings journal published by IOP Publishing.

- **Type of peer review:** Single Anonymous
- **Conference submission management system:** Morressier
- **Number of submissions received:** 13
- **Number of submissions sent for review:** 13
- **Number of submissions accepted:** 13
- **Acceptance Rate (Submissions Accepted / Submissions Received × 100):** 100
- **Average number of reviews per paper:** 1
- **Total number of reviewers involved:** 2
- **Contact person for queries:**
  Name: AMBROSE RAJKUMAR M
  Email: ambrose@kristujayanti.com
  Affiliation: KRISTU JAYANTI COLLEGE, BANGALORE, INDIA - Physics

IOP Conf. Series: Materials Science and Engineering 1233 (2022) 012001 doi:10.1088/1757-899X/1233/1/012001

Dosage optimisation of magnetically retrievable zinc ferrite nanoparticles for photodegradation of methylene blue

Minu Pius* **and Santhi Ani Joseph**

Department of Physics and Centre for Research, St. Teresa's College(Autonomous), Ernakulam, India-682011

* minupius@teresas.ac.in

Abstract. In this paper, we report the studies on photocatalytic activity of the zinc ferrite nanoparticles for the degradation of methylene blue dye. The study reveals that an optimum minimum dosage of 1g/L can give a degradation efficiency of 89% within 90 minutes of its exposure to the sunlight. Zinc ferrite nanoparticles were synthesised using co-precipitation method in stoichiometric ratio. Structural, magnetic and optical characterisation were done using XRD, FTIR, VSM and UV-VIS-IR analysis which confirmed the crystalline, magnetic and visible light active nature of the sample. Feasibility for magnetic retrieval of the photocatalyst was studied after photodegradation of dye and then analysed using XRD which confirmed the phase purity of the sample even after retrieval.

Keywords: magnetic photocatalyst, zinc ferrite, retrievability, methylene blue, dosage

1. Introduction

Synthetic dyes are widely used in pharmaceuticals, textiles, food manufacturing, cosmetics, plastics, paper and pulp industries. The discharge of these dyes and unreacted chemicals into the water bodies is a mutagenic and carcinogenic threat to the aquatic flora and fauna as well as for the animals and human beings. Some of these dyes due to their complex aromatic structure and synthetic origin are difficult to degrade(1–6).

Chemical, physical and biological processes are usually employed for the removal of these pollutants before discharge into the water bodies. But these methods either leads to toxic intermediates or secondary pollutants. Advanced oxidation processes (AOPs) can be used to degrade pollutants in aqueous media by strong oxidants such as hydroxyl radicals (OH). Homogeneous (Fenton reaction) or heterogeneous catalysts (metal catalyst, metal oxide, grapheme) are used to produce hydroxyl radicals from H_2O_2(7,8).

Photocatalytic processes are environmental friendly methods that utilize radiation energy to perform catalysis under ambient conditions and thus appear as the most emerging technology. Among these, visible light active photocatalysts are highly sought out as it make use of abundant sunlight, a clean energy, for photodegradtion of toxic dyes. Recently ferrites have gained importance due to low band gap and hence good visible light solar radiation absorbance, large availability of catalytic sites for adsorption of noxious waste/dyes, easy recovery after photocatalytic reaction due to the magnetic nature and high stability.

The present work aims at dosage analysis of zinc ferrite nanoparticles as photocatalysts in the degradation of methylene blue dye and subsequent magnetic retrieval from the solution.

2. Experimental Details

2.1 Synthesis

Zinc Ferrite is synthesized via co-precipitation method using ferric chloride hexahydrate and zinc chloride precursors and sodium hydroxide as precipitating agent in the stoichiometric ratio. The precipitate obtained is kept in a furnace for 2 hours at digestion temperature of 100^0c. The precipitate is then washed and centrifuged at 6000rpm subsequently for several times. The precipitate thus collected is dried at 100°C for 4 hours. Dried powder is grounded using agate mortar and pestle. The powder is then calcined at 600°C for 6 hours to obtain the crystallized zinc ferrite nanomaterial, labelled as ZF.

2.2 Characterisation

The purity and crystalline nature of as-synthesised samples are studied using Bruker AXS D8 Advance X-ray diffractometer with Cu-Kα radiation (λ = 1.54 A^0) and Fourier Transform Infrared Spectrometer (Bruker-Alpha-Platinum) in the range 4000-500cm^{-1}. Absorption spectrum and diffusive reflectance spectrum are obtained using Perkin Elmer lambda 950 UV-VIS-NIR instrument in the range 400-1500nm. Magnetization measurements of samples at room temperature are carried out on a Vibrating Sample Magnetometer (Lakeshore VSM 7410) with an applied field -15 kOe < H < 15 kOe.

Photocatalytic degradation of methylene blue dye(10ppm) is tested using different dosages such as 0.5g/L , 1g/L and 1.67g/L of sample as the photocatalyst in solar irradiation for 200 minutes. Required aliquots are taken at regular intervals and analysed by taking UV-VIS-NIR absorption spectrum. The efficacy in degradation is then analysed for the optimum dosage with an initial stirring for 20 minutes at dark followed by solar irradiation for 90 minutes. The photocatalysts are then retrieved using permanent magnets and stability of the sample is checked using X-ray diffraction analysis.

3. Results and Discussion

3.1 X-Ray diffraction Studies

The X-ray diffraction from various lattice planes of sample (ZF), shown in figure 1(a), reveals the phase purity and crystalline nature of the synthesised sample. Diffraction peaks obtained at 2θ values of 29.92^0, 35.265^0, 42.84^0, 53.1^0, 56.6^0, 62.2^0 and 73.515^0 are indexed to the lattice planes of (hkl) values (220), (311), (400), (422), (511), (440) and (533) respectively. This is in agreement with standard JCPDS File No: 00-022-1012 confirming cubic lattice of spinel phase with Fd-3m(227) space group.

The lattice parameter a is evaluated to be 8.44A^0 from the most intense peak (311) of $ZnFe_2O_4$ using the equation (1) and matches well with the standard value of 8.441A^0, for cubic crystal structure of spinel ferrite.

$$a = d\sqrt{h^2 + k^2 + l^2} \quad (1)$$

The crystalline size (D), determined using Debye Scherrer equation given by equation (2), for the most intense peak in the XRD pattern (311), is found to be 13nm.

$$D = \frac{0.9\lambda}{\beta \cos\theta} \quad (2)$$

IOP Conf. Series: Materials Science and Engineering 1233 (2022) 012001 doi:10.1088/1757-899X/1233/1/012001

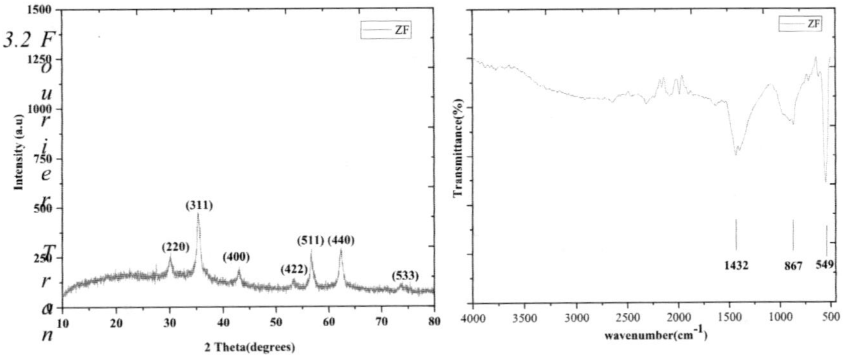

Figure 1. (a)XRD pattern of zinc ferrite nanoparticles. (b) FTIR spectrum of zinc ferrite nanoparticles.

3.2 Fourier Transform Infra-red Analysis

The presence of characteristic peaks of metal-oxide vibration in FTIR spectrum shown in figure 1(b) confirms the spinel ferrite structure of the sample(9). These peaks at 549cm^{-1} and 867cm^{-1} are assigned to the metal-oxide (Zn-O) stretching vibration in tetrahedral site and Fe-O stretching vibration in octahedral site respectively. The absorptions at 1432cm^{-1} and 1402cm^{-1} is attributed to the symmetric CH_2 bending vibration. Feeble absorption in the broad range of 3400-2500cm^{-1} and at 1600cm^{-1} corresponding to hydroxyl groups and overtone respectively confirms that the sample contain very little presence of any adsorbed water molecules.

3.3 UV-VIS-IR absorption spectrum

The absorption spectrum in figure 2(a) shows the broad range visible light active nature of the sample. α value is evaluated using Kubelka Munk relation given in equation (3) for the powdered sample connecting scattering (s) and absorption coefficient (k) with diffused reflectance (R).

$$\alpha = \frac{k}{s} = \frac{(1-R)^2}{2R} \tag{3}$$

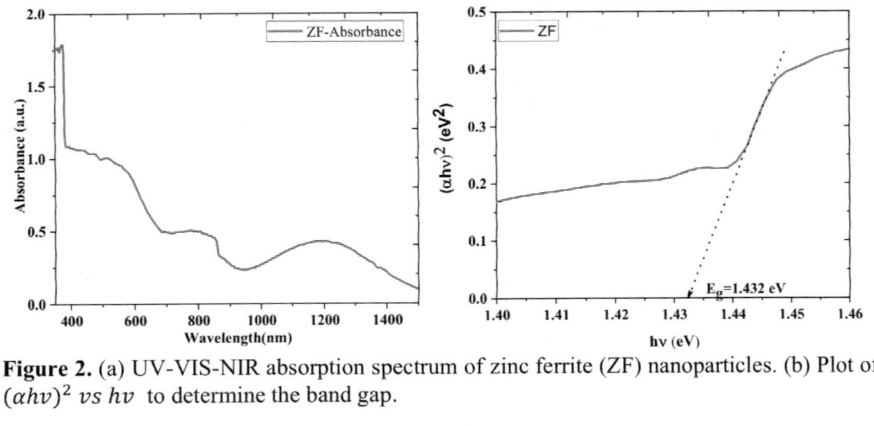

Figure 2. (a) UV-VIS-NIR absorption spectrum of zinc ferrite (ZF) nanoparticles. (b) Plot of $(\alpha h\nu)^2$ vs $h\nu$ to determine the band gap.

The linear region of the plot of $(\alpha hv)^2$ vs hv is extrapolated to get a band gap of 1.432eV as shown in figure 2(b) which is lesser that the standard value of 1.92eV. This red shift can be attributed to additional sub-band-gap energy levels introduced by point defects, well in line with the previously reported observations that surface and interface defects in zinc ferrite nanoparticles may cause a redshift of band gap(10). Thus the red shift enhances visible light activation of the sample for photoreaction.

3.4 Vibration Sample Magnetometer Studies

The S-shaped M-H curve as shown in figure 3 reveals the superparamagnetic nature of the sample. The saturation magnetisation (Ms) is found out to be 6.17emu/g and a remanence (Mr) as 7.65×10^{-5}emu/g gives a low squareness ratio(Mr/Ms) of 1.23×10^{-5} which indicates a better single domain formation and superparamagnetic behaviour.

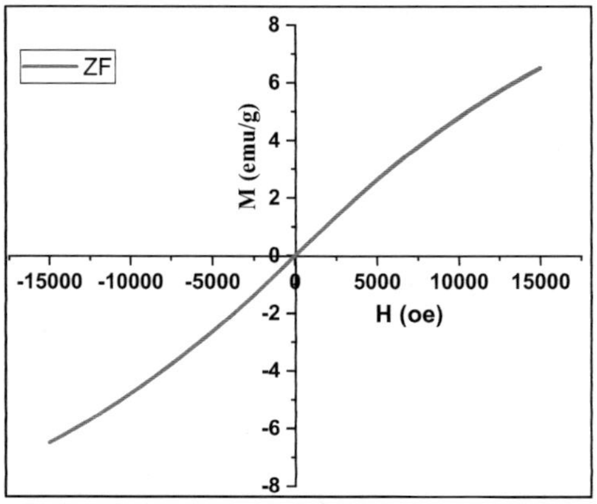

Figure 3. M-H curve of zinc ferrite nanoparticle.

3.5 Photocatalytic Studies

The absorption spectra of methylene blue (MB) dye containing different dosages of 0.5g/L (ZF 30), 1g/L (ZF 60) and 1.67g/L (ZF 100) of zinc ferrite as photocatalyst taken at different time intervals of solar irradiation are shown in figure 4(a)-(c). The maximum peak at a wavelength of 664 nm is due to π conjugation between two dimethylamine substituted aromatic rings through S and N and the other absorption at 292 and 245 nm is related to the phenothiazine species. According to Beer–Lamberts' law, the concentration of MB is proportional to its absorbance. Thus a reduction in the peak absorbance with time indicates a reduction in the concentration of dye. Degradation efficiency(%) is evaluated for each case using the equation (4) where C_0 is initial dye concentration and C_t is dye concentration at different times(11) and listed in table 1.

$$\% \text{ Degradation}= \frac{C_0-C_t}{C_0} \times 100 \quad (4)$$

% Degradation of MB with different dosages shown in figure 4(d) indicates higher degradation of dye occurs for both 1g/L (ZF 60) and 1.67g/L (ZF 100) dosages than in 0.5g/L(ZF 30) in 200 minutes. Though faster degradation is seen for 1.67g/L, 1g/L dosage is chosen for further analysis as both gives same amount of degradation after 200 minutes and minimum of photocatalyst is hence preferred for photocatalysis.

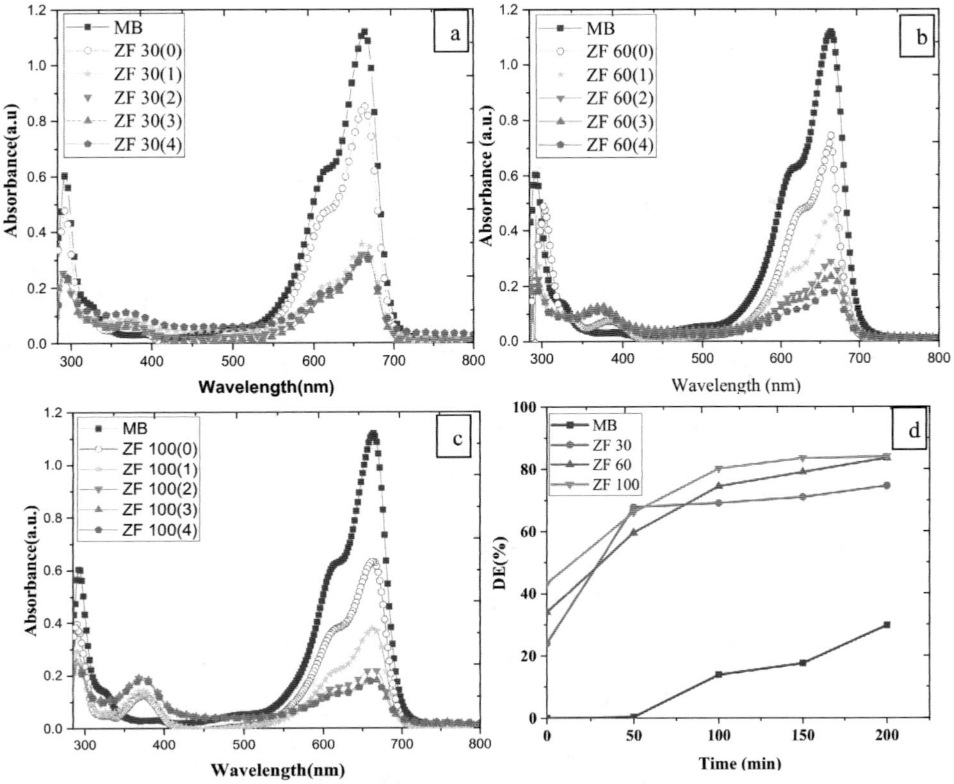

Figure 4. (a)-(c)Absorption spectrum of methylene blue dye after photo degradation in the presence of different dosages of zinc ferrite nanoparticles in the order 0.5g/L(ZF 30), 1g/L (ZF 60), 1.67g/L (ZF 100) at exposures 0f 50 minutes each (d)Degradation efficiency of dye without photocatalyst as MB and with zinc ferrites as ZF 30, ZF 60 and ZF100 .

Table 1. Degradation efficiency of dye with and without photocatalyst.

Time	Degradation Efficiency(%)			
min	**MB**	**ZF 30**	**ZF 60**	**ZF 100**
	10ppm	**$0.5gL^{-1}$**	**$1gL^{-1}$**	**$1.67gL^{-1}$**
0	0	24	34	43.4
50	0.33	67.8	59.6	66.2
100	13.8	69	74.4	80
150	17.5	71	79	83.4
200	29.7	74.6	83.5	84

Optimised dosage of 1g/L of sample in methylene blue dye solution is further analysed. After addition of photocatalyst, the dye solution is stirred in dark for 20 minutes for attaining equilibrium adsorption of the dye molecules onto the surface and then it is exposed to sunlight for 90 minutes. There is significant enhancement in the rate of degradation as shown in figure 5(a)-(c). 85% of degradation was obtained in a faster duration of 90 minutes.

The degradation efficiency demonstrated by zinc ferrite sample in the present work is significant compared to the highest reported value of 70% in 5 hours for 6.4ppm MB dye using 1g/L of ZF (12) and those reported in (13) using 15nm zinc ferrite (1g/L) and H_2O_2 synthesised using Hydrothermal method showed only 52% in 6 hours. Higher degradation are reported for zinc ferrite in the presence of H_2O_2 at higher dosages or in composite form with MWNT, graphene,β- $AgVO_3$ at longer exposure time(14–16).

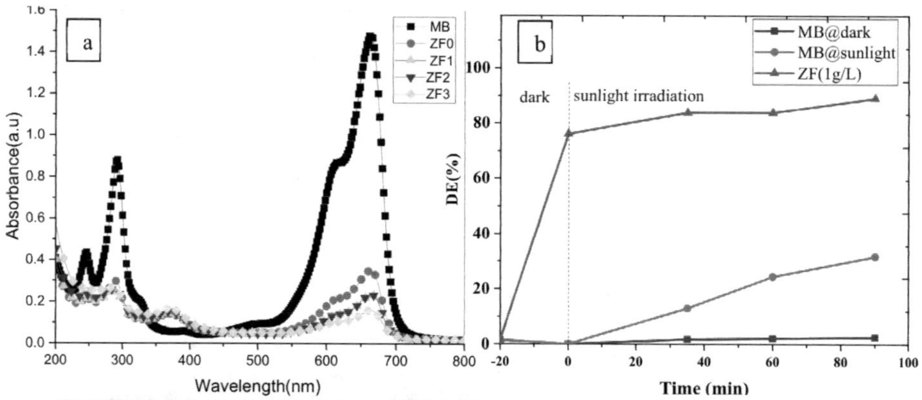

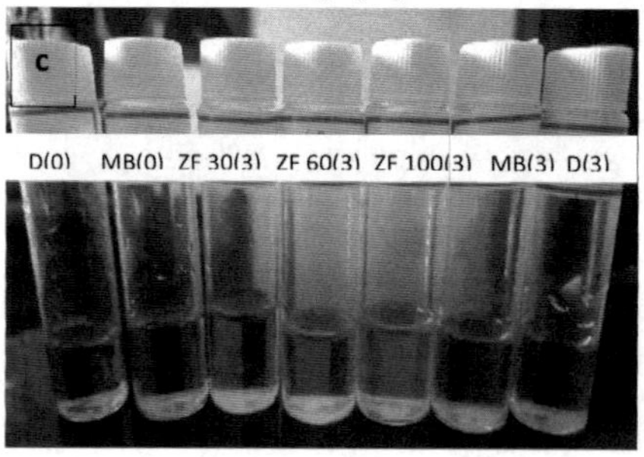

Figure 5. (a) Absorbance spectrum of MB dye after photodegradtion in the presence of 1g/L of zinc ferrite in a time interval of 30 minutes each. (b) % Degradation of dye initially at dark for 20 minutes and then exposed to sunlight.(c) Photograph of dye kept in dark at 0 min D(0), dye in sunlight MB at 0 min MB(0), ZF30(3), ZF 60(3), ZF 100(3), dye in sunlight MB(3), dye in dark D(3) after 90 min

3.6 *Retrievability of Photocatalyst*
As the zinc ferrite are superparamagnetic, it was retrieved from the dye solution after photodegradtion. The samples were then washed, dried and analysed using XRD studied. Diffraction peaks obtained in XRD shown in figure 6, are clearly indexed with the standard peaks of zinc ferrite indicating the stability of the sample after photodegradation.

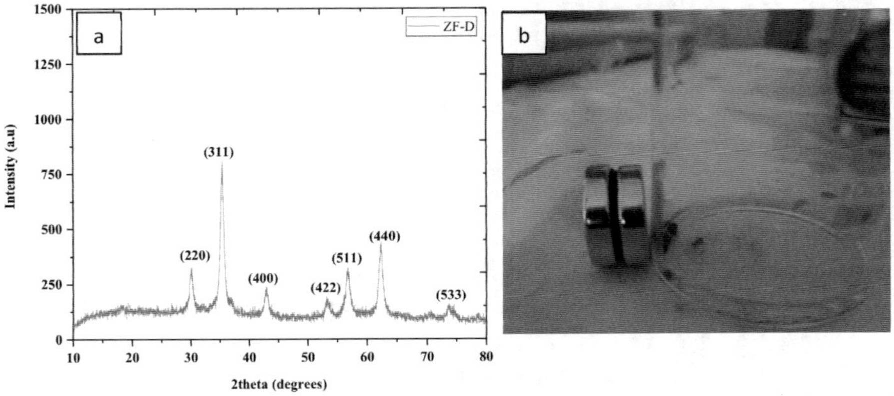

Figure 6. a)XRD pattern of zinc ferrite retrieved after photodegradation of dye. b) Magnetic retrieval of sample from the dye solution after photodegradation

4 Conclusion
Zinc ferrite nanomaterials are synthesised using co-precipitation method and magnetic, optical and structural characterisation of the synthesised sample confirmed the superparamagnetic, visible light active and cubic lattice of spinel ferrite formation with a crystallite size of 13nm. Photodegradation of 10ppm methylene blue under sunlight irradiation are obtained within 200 minutes with a degradation efficiency of 69%, 85% for 0.5g/L, 1g/L and 1.67g/L dosage of zinc ferrite as photocatalyst respectively. Initial stirring of optimised dosage (1g/L) in the dye solution at dark facilitated equilibrium adsorption of dye molecules and increased rate of degradation within 90 minutes of exposure to sunlight. Magnetic retrievability of the photocatalyst is checked and stability of the sample is confirmed using XRD analysis.

Acknowledgement
The authors wish to acknowledge MARG fund instituted by St. Teresa's College (Autonomous), Ernakulam and the help rendered by the Department of Physics and Zoology of St. Teresa's College, Ernakulam, SAIF IIT Madras and STIC CUSAT for the characterisation of the sample.

References
1. Mahmoodi NM. Photocatalytic ozonation of dyes using copper ferrite nanoparticle prepared by co-precipitation method. Desalination. 2011 Sep;279(1–3):332–7.

2. Zainal Z, Hui LK, Hussein MZ, Abdullah AH, Hamadneh I (Moh'd KR. Characterization of TiO2–Chitosan/Glass photocatalyst for the removal of a monoazo dye via photodegradation–adsorption process. J Hazard Mater. 2009 May;164(1):138–45.

3. Özcan A, Özcan AS. Adsorption of Acid Red 57 from aqueous solutions onto surfactant-modified sepiolite. J Hazard Mater. 2005 Oct;125(1–3):252–9.

4. Fan H, Shu H, Tajima K. Decolorization of acid black 24 by the FeGAC/H2O2 process. J Hazard Mater. 2006 Feb 6;128(2–3):192–200.

5. Santos VP, Pereira MFR, Faria PCC, Órfão JJM. Decolourisation of dye solutions by oxidation with H2O2 in the presence of modified activated carbons. J Hazard Mater. 2009 Mar;162(2–3):736–42.

6. Mahmoodi NM, Najafi F. Preparation of surface modified zinc oxide nanoparticle with high capacity dye removal ability. Mater Res Bull. 2012 Jul;47(7):1800–9.

7. Mahmoodi NM. Photocatalytic ozonation of dyes using multiwalled carbon nanotube. J Mol Catal Chem. 2013 Jan;366:254–60.

8. Kurniawan TA, Lo W. Removal of refractory compounds from stabilized landfill leachate using an integrated H2O2 oxidation and granular activated carbon (GAC) adsorption treatment. Water Res. 2009 Sep;43(16):4079–91.

9. Waldron RD. Infrared Spectra of Ferrites. Phys Rev. 1955 Sep 15;99(6):1727–35.

10. Harish KN, Bhojya Naik HS, Prashanth kumar PN, Viswanath R. Synthesis, enhanced optical and photocatalytic study of Cd–Zn ferrites under sunlight. Catal Sci Technol. 2012;2(5):1033.

11. Rashad MM, Mohamed RM, Ibrahim MA, Ismail LFM, Abdel-Aal EA. Magnetic and catalytic properties of cubic copper ferrite nanopowders synthesized from secondary resources. Adv Powder Technol. 2012 May;23(3):315–23.

12. Arimi A, Megatif L, Granone LI, Dillert R, Bahnemann DW. Visible-light photocatalytic activity of zinc ferrites. J Photochem Photobiol Chem. 2018 Nov;366:118–26.

13. Chen C-H, Liang Y-H, Zhang W-D. ZnFe2O4/MWCNTs composite with enhanced photocatalytic activity under visible-light irradiation. J Alloys Compd. 2010 Jul;501(1):168–72.

14. Lu D, Zhang Y, Lin S, Wang L, Wang C. Synthesis of magnetic ZnFe2O4/graphene composite and its application in photocatalytic degradation of dyes. J Alloys Compd. 2013 Dec;579:336–42.

15. Chen P. Synthesis and photocatalysis of novel magnetic reduced graphene oxide-ZnFe2O4 nanocomposites with highly efficient interface-induced effect. J Sol-Gel Sci Technol. 2017 May;82(2):397–406.

16. Jenita Rani G, Jothi Rajan MA, Gnana kumar G. Reduced graphene oxide/ZnFe2O4 nanocomposite as an efficient catalyst for the photocatalytic degradation of methylene blue dye. Res Chem Intermed. 2017 Apr;43(4):2669–90.

Thermal and structural studies of Bismuth Borate glasses modified with Gadolinium Oxide for solid state lighting applications

M Prasanth[1], N Ravi[2*], K Thyagarajan[3] and K Venkata Krishnaiah[2]

[1] Department of Physics, Jawaharlal Nehru Technological University Anantapur, Anantapuramu, AP -515002, India
[2] Department of Physics, Rajeev Gandhi Memorial College of Engineering and Technology, Nandyal, AP -518501, India
[3] Department of Physics, Jawaharlal Nehru Technological University Anantapur College of Engineering, Pulivendhula, AP -516390, India

Abstract. Bismuth based borate glasses with combination B_2O_3 –Bi_2O_3 – Gd_2O_3 - K_2O- Li_2O (BBiGdKLi) were prepared by melt-quenching method at 1250 °C for thermal and structural properties investigation. . Glass transition temperature (T_g) was observed at 681 K in heat capacity (C_p) vs temperature plot. As the temperature increases the specific heat or heat capacity also increases but a sudden fall in C_p was noticed at 962 K which is known to be crystalline temperature (T_c) of BBiGdKLi glass. Borate units comprising non-bridging oxygen (NBO) such as υ(B-O-Gd/Bi/K/Li) occured due to asymmetric stretching vibration at 976 cm^{-1} revealed in Raman spectrum excited by 532 nm laser. B1s, Bi 4f and Gd 4d binding energies from the XPS spectra of BBiGdKLi glass are corresponding to 188.9, 156.35 & 161.75 and 156.35 & 161.7 eV, respectively. The thermal behavior of BBiGdKLi glass suggest that these glasses could be significant materials for solid state lighting applications.

1. Introduction

Among all the inorganic binary and ternary glasses, the borate system is diverse, including excellent glass-forming ability, high transparency, relatively low melting temperature, high thermal stability and good solubility for rare earth (RE) ions. Alkaline metal borate glass at the top resists atmospheric moisture and receives a large number of transition metals with transparency ranging from almost ultraviolet to visible and medium infrared [3]. Lithium (Li) as a glass modifier can reduce the hygroscopicity of the glass and stabilize the glass by introducing non-bridging oxygen (NBO) that forms the color center. Strontium (Sr) was created to improve the optical and structural properties of modified lithium borate glass [4]. However, borate glass without modifiers has high acoustic energy (1300 cm^{-1}), can cause non-radiation and reduce the intensity of RE ion emissions [5]. Therefore, the addition of modifiers to borate glass is necessary to improve its physical and spectral properties. The addition of sodium oxide to borate changes its structural units and the coordination of boron, which improves its physical and spectral properties [6]. Noble glass formers are required to develop efficient lasers from UV, visible, near IR, mid IR wavelength region.

Glass containing Bi_2O_3 has a high refractive index and this property is very important for advanced optical communication and processing devices [7]. Much research has been done on the physical properties of glass, including Bi_2O_3, but there is still much controversy about the relationship between structures at the individual level and their macroscopic properties. Mogus-Milankov et al., [8] Bi_2O_3-

IOP Conf. Series: Materials Science and Engineering 1233 (2022) 012002 doi:10.1088/1757-899X/1233/1/012002

containing glass is considered to be composed of highly deformable BiO_6 groups, as previously proposed by Bishay and Maghrabi, while other authors assume that BiO_3 is a former group, on the other hand, the BiO_6 group is a modifier. Contemporary authors claim that both BiO_6 and BiO_3 units are present in glass containing Bi_2O_3 and that the proportion of each unit in these groups depends on the ratio of oxides to other glass components [9].

Owing to high refractive index and non-linear optical properties of heavy metal borate glass doped with RE elements are particularly useful for solid state lasers. Gadolinium oxide (Gd_2O_3) provides a platform for co-doping with other RE ions due to its density (ρ = 7.41 g/cm^3) and the active energy transfer (ET) properties of the fused glass [10].

This article presents a series of bismuth borate glass with 40 B_2O_3 -30 Bi_2O_3 -10 Gd_2O_3 -10K_2O -10Li_2O (BBiGdKLi) using melt quenching system for structural and thermal studies. To assess the thermal behaviour of the glass, the thermal heat capacity or specific heat related to temperature was analysed. The Raman spectrum was used to analyse the bond between the added element and the vibrating glass formation zone and the binding energies of B, Bi and Gd were evaluated using the XPS spectrum.

2. Experimental procedure

The bismuth oxide containing bismuth borate glass (BBiGdKLi) was prepared by melt quenching technique. All of these chemicals are grades of reagents purchased from Alfa Aesar. A batch of 15 g was placed in agate mortar and pestle and ground for 2 hours to give a fine, homogeneous powder. In addition, this fine homogeneous powder was placed in an alumina crucible and melted at 1250 °C for 2 hours using an electric oven. After some time, the melt was abruptly removed from the furnace and quenched in air on a preheated brass plate at room temperature (RT). In addition, these samples were annealed at 350 °C for 12 hours and cooled to room temperature. These samples are polished for optical quality and used for optical and structural characterization.

3. Characterization techniques

FTIR spectra were recorded for the glass powder sample using Jasco FTIR-4100 in the range from 400 to 2500 cm^{-1}. Photoluminescence spectra (visible up-conversion and NIR emission) and decay curves were obtained by 980 nm laser excitation in the continuous and pulse modes, respectively, by the use of visible as well as NIR detectors connected Edinburg spectrofluorometer (FLS-980) at room temperature.

4. Results and analysis

4.1 Thermal analysis

Glasses and crystals of heat-capacity differences have revealed much intension due to their behavior at low temperatures. In general, from the findings the heat capacity of the glass is greater compared to the crystal and could be find that the heat-capacity difference between them reaches to a maximum value. This maximum and large value of the heat-capacity difference at low temperature are ascribed to the presence of low-frequency modes in the glass those are not exhibited by the crystals [11]. Amorphous materials contain an extra vibrational density of states (VDOS) labelled as $g(v)$, compared to prophesied by the Debye theory in the terahertz frequency range. This additional VDOS establishes as a larger heat capacity (C_p) compared to Debye T^3 law in the low-temperature region where a bump called 'boson peak' come out in the C_p /T^3 vs T plot. This boson peak also seen in the low-energy region as a broad band of Raman/inelastic-neutron scattering spectra which states a common feature in amorphous matter owing to the boson peak that araise from the inorganic, organic, and metal substances [12, 13].

Fig. 1. displays change of C_p with temperature (K) BBiGdKLi glass which was recorded in the range from 200 -1400 K. At a low temperature of 321 K, BBiGdKLi glass shows a *Cp* value, which increases rapidly with a small increase in temperature. At temperatures up to 357 K. Glass transition temperature (T_g) was observed at 681 K. Later, the linear behaviour of *Cp* in BBiGdKLi glass with temperature was reported. After heat treatment at 350 °C for 12 hours, a sharp drop in *Cp* at 962 K was observed, representing the crystalline temperature (*Tc*) of BBiGdKLi glass. The melting temperature of glass (T_m) is 1067 K.

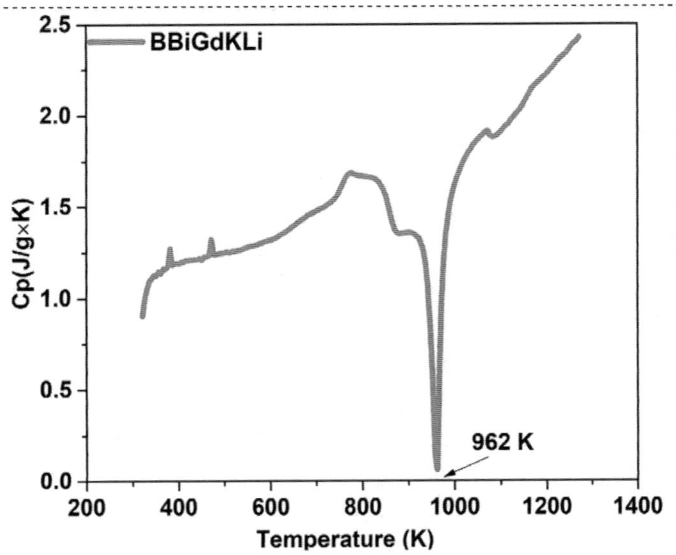

Figure. 1. Temperature vs heat capacity of BBiGdKLi glass.

4.2 Raman analysis

The Raman spectrum of BBiGdKLi was recorded in the range 200-1200 cm^{-1} and excited ions with a laser wavelength of 532 nm, as shown in Fig. 2. The spectrum shows a wider and stronger band observed at about 550 cm $^{-1}$. This is due to the bending function υ (BO_4). Due to the bending of B-O-B bond in the borate lattice, a band of about 755 cm^{-1} appeared in all spectra. Raman bands approximately at 850 cm^{-1} and 990 cm^{-1} in borate-based glass are due to the large number of non-briding oxygen (NBO) forming borate units such as the u band (B-O-Gd/Bi/K/Li). This range is due to asymmetric stretching vibrations at 976 cm-1. Raman studies have clearly shown that the presence of MX_2 (X = Cl, Br, F) increases the intensity of these vibrational bands associated with BO_4 units and NBOs. These are assigned to pentaborate groups [14].

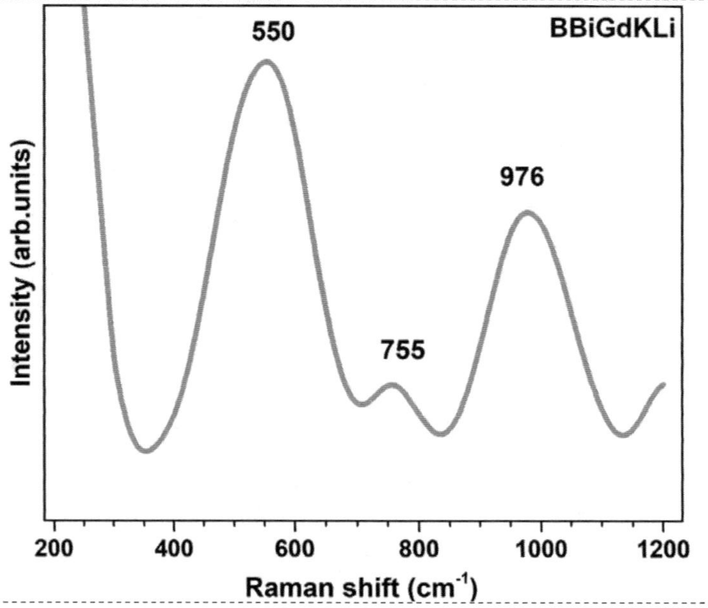

Figure. 2. Raman spectra of BBiGdKLi glass excited by 532 nm laser.

4.3 XPS spectra (B, Bi and Gd)

BBiGdKLi, a bismuth borate glass, is mainly associated with an ancient borate lattice structure consisting of three coordinated boron ($BO_{3/2}$) and four coordinated boron ($BO_{4/2}$). In contrast, the oxygen bound to boron cation (Gd/Bi/K/Li) is called NBO and briding oxygen (BO) in the matrix. The oxygen binding energy (BE) in the X-ray photoelectron spectroscopy (XPS) is the oxygen state and distinguishes between BO and NBO. In addition, a peak at 188.9 eV as exhibited in Fig. 3(a) is in the BBiGdKLi glass of B*1s* electron is attributed to H_3BO_3 or B_2O_3 [15].

The BE of electrons in a material is determined by the level at which the emission of photoelectrons occurs in two main factors. (i) the state of atomic oxidation and (ii) the local microscopic chemical and physical environment. Changes between these parameters lead to subtle changes to the position of the peaks in the XPS range. This is characterized as a chemical change. These changes are superficial and easily explained in the XPS spectra. These different oxidation states and the chemical environment are one of the major advantages of XPS technology. In the fabricated BBiGdKLi glass, two strong Bi ion peaks were resolved at 161.75 eV (Bi $4f_{5/2}$) and 156.35 eV (Bi $4f_{7/2}$) as shown in Fig. 3(b). XPS spectra of Gd 4*d* electrons exist in the BBiGdKLi glass is displayed in Fig. 3(c). Gd 4*d* electrons are resolved in two valence states the peaks located with the BE values around 161.7 eV ($4d_{3/2}$) and 156.35 eV ($4d_{5/2}$).

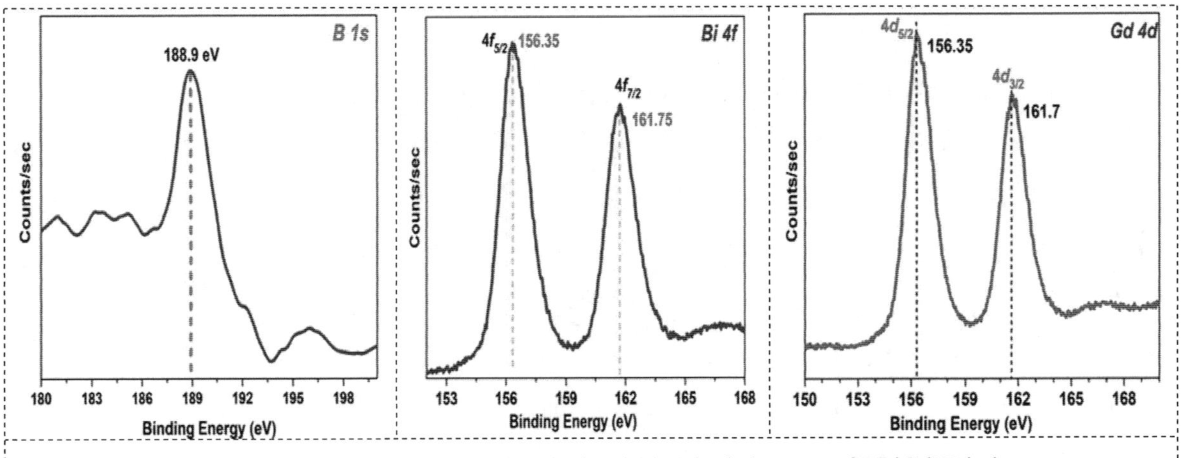

Figure. 3. XPS spectra of (a) B 1s, (b) Bi 4f and (c) Gd 4d elements of BBiGdKLi glasses

5. Conclusion

Bismuth Borate-BBiGdKLi glass was fabricated using the common melt quenching method to evaluate the thermal and structural properties. Melting temperature, crystal temperature and glass melting temperature were estimated from the thermal properties of the BBiGdKLi glass using a heat capacity versus a temperature curve. The crystalline temperature Tc of the BBiGdKLi glass was estimated and found to be 962 K. Raman spectra of BBiGdKLi glass revealed bridging and non-bridging oxygen bonds to the borate network. We evaluated the binding energies of B *1s*, Bi *4f* and Gd *4d* electrons in BBiGdKLi glass. Bi^{3+} and Gd^{3+} ions as heavy metal ions enhance the optical properties of BBiGsKLi glass for light source. The *Cp* of BBiGdKLi glasses may indicate that these glasses are useful for lighting applications due to their thermal stability.

6. References

[1] I Kashif, A Abd El-Maboud, and A Ratep 2014 *Results Phys.* **4** 1–5.
[2] M H A Mhareb, S Hashim, S K Ghoshal, Y S M Alajerami, M A Saleh, R S Dawaud, N A B Razak, S A B Azizan. 2014 *Opt. Mater.* (Amst). **37** 391–397.
[3] P Venkat Reddy, C Laxmi Kanth, V Prashanth Kumar, N Veeraiah, and P Kistaiah 2005 *J. Non. Cryst. Solids*, **351** 3752–3759.
[4] H K Obayes, H Wagiran, R Hussin, and M A Saeed 2016 *J. Mol. Struct.* **1111** 132–141.
[5] Y.G. Choi, J. Heo, 1997 *J. Non- Cryst. Solids* **217** 199–207.
[6] T Yano, N Kunimine, S. Shibata, M. Yamane 2003 *J. Non- Cryst. Solids* **321** 147–156.
[7] Sugimito N, Ultrafast Optical Switches and Wavelength Division Multiplexing (WDM) Amplifiers Based on Bismuth Oxide Glasses 2002 *J. Am. Ceram. Soc.* **85** (5) 1083–88.
[8] ElBatal FH 2007 *Nucl Instr and Meth Phys Res (B)* **254** (2) 243-253.
[9] F H ElBatal, M A Marzouk, A M Abdel ghany 2011 *J. Mater. Sci.* **46** 5140–5152.
[10] Y Al-Hadeethi, M I Sayyed J Kaewkhao, Bahaaudin M Rafah Rahma Almalki, R Rajaramakrishna 2019 *Appl. Phys. A* **125** 749 1-10.
[11] Charles M Guttman 1972 *The J. of Chem. Phys.* **56** 627-630.
[12] Kensaku Nakamura, Yoshihiro Takahashi & Takumi Fujiwara 2014 *Scientific Reports* **4** 6523.
[13] Safarik, D J, Schwarz, R B & Hundley, M F 2006 *Phys. Rev. Lett.* **96** 195902–195905.
[14] Jiaoyu Peng, Jing Chen, Yaping Dong, Wu Li 2018 Spectrochim. Acta Part A: Mole. and Biomole. Spec. **199** 367–375.
[15] Miguel Quesada-Gonzalez, Nicolas D Boscher, Claire J Carmalt, and Ivan P Parkin 2016 *ACS Appl. Mater. Interfaces* **8** 38 25024–25029.

Temperature dependent magnetic studies of $Ba_{0.5}Sr_{0.5}Co_{0.2}Fe_{11.8}O_{19}$ prepared by solid state reaction method

V Argish[1], M G Shalini[1], Subasa C Sahoo[1]

[1]Department of Physics, Central University of Kerala, Kasaragod, Kerala 671320, India

Email: subasa.cs@gmail.com

Abstract. $Ba_{0.5}Sr_{0.5}Co_{0.2}Fe_{11.8}O_{19}$ was synthesized using solid state reaction method. Lattice parameters 'a' and 'c' were found to be 5.886 Å and 23.110 Å respectively. Scanning electron microscopy studies showed platelet type structures in the sample and the elemental mapping showed the uniform distribution of cations throughout the sample. The spontaneous magnetization (M_S) of 62 emu/g with coercivity (H_C) of 1.22 kOe was observed at 300K. The M_s value increased whereas the H_C decreased with the decreasing measurement temperature. The highest M_s value of 88 emu/g was observed at 60K. Curie temperature of 819 K was observed for the sample. The anisotropy constant (K_1) and anisotropy field (H_a) were found to be increased with the decrease in measurement temperature. The observed magnetic behaviour can be explained on the basis of distribution of cations in the lattice sites and shape anisotropy of the prepared sample.

1. Introduction

M-type hexagonal ferrites like $BaFe_{12}O_{19}$ and $SrFe_{12}O_{19}$ play major role in the manufacturing of permanent magnets and magneto-optical recording devices due to their high chemical stability, high magneto crystalline anisotropy, high saturation magnetization, high coercivity and high Curie temperature [1-2]. The unit cell of M-type hexagonal ferrite $BaFe_{12}O_{19}$ consists of two molecules in SRS*R* stacking, where * indicates a 180° rotation of the block around the c-axis. R and R* are hexagonal blocks; S and S* are spinel blocks. Ba^{2+} ions have no unpaired electrons and Fe^{3+} ions having 5 unpaired electrons are the magnetic ions in the unit cell. The 24 Fe^{3+} ions are situated in tetrahedral site (4f1), octahedral sites (2a,4f2 and 12k) and trigonal bipyramidal site (2b) in the unit cell. The properties of these hexagonal ferrites depend on the synthesis techniques, sintering temperature and substitution of different ions. Different synthesis techniques like sol-gel method, co-precipitation method, ceramic method, hydrothermal method have been adopted to synthesize these hexaferrites. In the present study, we synthesized $Ba_{0.5}Sr_{0.5}Co_{0.2}Fe_{11.8}O_{19}$ by ceramic method and studied its structural, microstructural and temperature dependent magnetic properties.

2. Experimental Details

Standard solid state reaction method was used for the preparation of $Ba_{0.5}Sr_{0.5}Co_{0.2}Fe_{11.8}O_{19}$. Stochiometric amount of $BaCO_3$ (purity 99%; Alfa Aesar), $SrCO_3$ (purity ≥ 98%;Loba Chemie, India), Co_3O_4 (purity ≥ 99%; SRL, India) and α-Fe_2O_3 (purity ≥ 99%; SRL, India) were mixed and ground for 3hrs using an agate mortar pestle. The mixture was calcined at 1000°C for 8hrs in air. The powder was ground again and calcined at 1200 °C for 8hrs. Then a pellet was made of the calcined powder. To make pellet, 2-4 drops of poly vinyl alcohol (PVA) solution was added and a load of 5 ton was used. The pellet was finally sintered at 1200 °C for 8hrs. The X-ray diffraction (XRD) data of the sample was taken between 20 to 80° in θ/2θ mode by a Rigaku SmartLab X-Ray Diffractometer with Cu-Kα radiation (λ=1.5406 Å). Raman spectra were recorded at room temperature in the range of 100 - 900 cm^{-1} using a Horiba LabRAM HR Raman spectrometer equipped with a solid state laser of wavelength 532nm. Fourier transformed infrared (FTIR) spectra were recorded using a Perkin Elmer Frontier FTIR spectrometer at room temperature in the range of 400 - 4000 cm^{-1}. The morphology and microstructure of the sample were studied by Ziess Ultra 55 scanning electron microscope (SEM). The magnetic studies were performed by a Quantum Design Versalab PPMS by applying ± 30 kOe field at different temperatures ranging from 310 K to 50 K.

3. Results and discussions

3.1. Structural and micro-structural studies

XRD pattern of the sample is shown in the Figure 1(a). It is seen in the Figure 1(a) that the sample is single phase. All the peaks were indexed to $Ba_{0.5}Sr_{0.5}Co_{0.2}Fe_{11.8}O_{19}$ (JCPDS card no: 00-051-1879 for $Ba_{0.5}Sr_{0.5}Fe_{12}O_{19}$). The lattice parameters 'a' and 'c' were found to be 5.89Å and 23.11Å respectively.

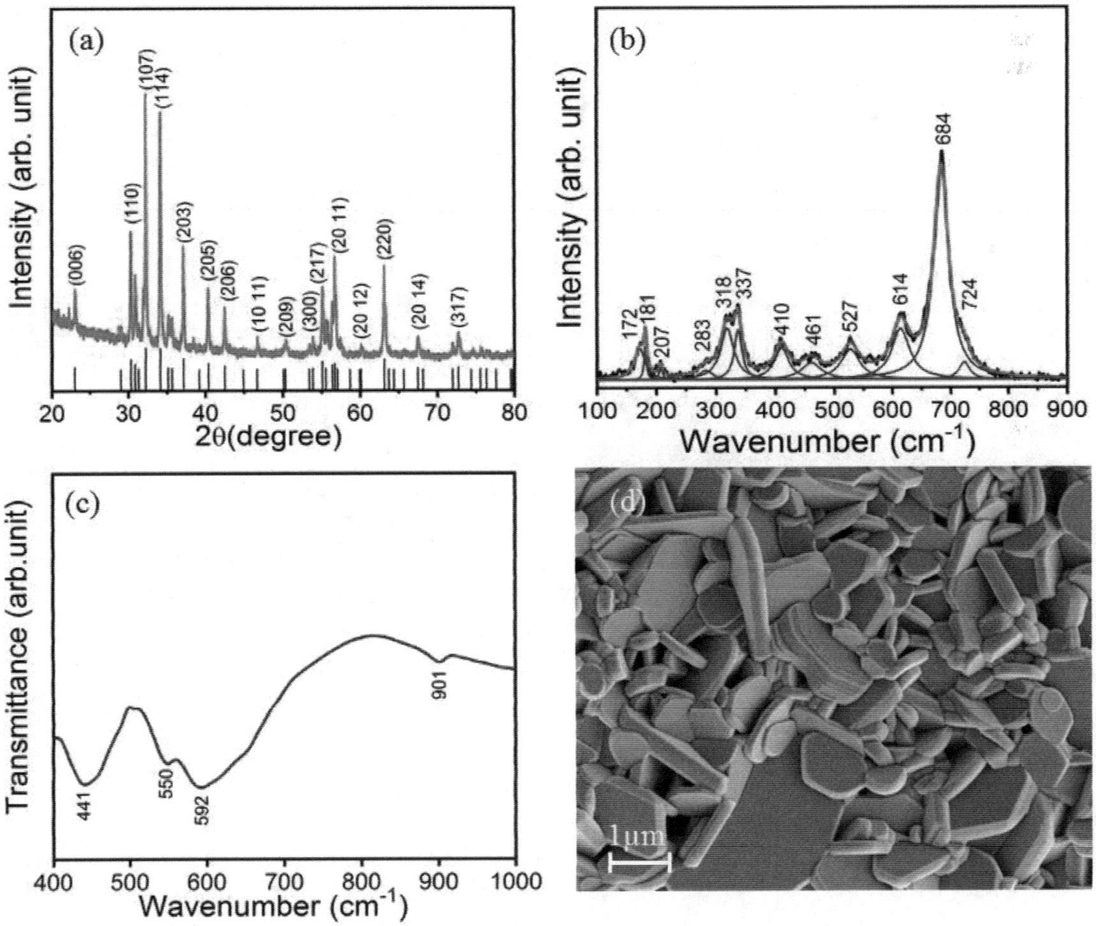

Figure 1. (a) XRD pattern, (b) Raman Spectra, (c) FTIR spectra and (d) SEM micrograph of $Ba_{0.5}Sr_{0.5}Co_{0.2}Fe_{11.8}O_{19}$. (Diffraction pattern of standard $Ba_{0.5}Sr_{0.5}Fe_{12}O_{19}$, JCPDS card no: 00-051-1879, is also shown below in Figure 1(a))

According to group theory 42 Raman active modes are predicted at room temperature for M-type hexagonal ferrite. In the present study, 12 Raman modes are observed in the range of 100 – 900 cm^{-1} as seen in Figure 1(b). The assignment of observed peaks for different symmetry and crystallographic sites are listed in Table 1. Observed Raman spectra are similar to those reported by others in M-type hexaferrite [3-7].

Table 1. Assignment of Raman active modes observed in $Ba_{0.5}Sr_{0.5}Co_{0.2}Fe_{11.8}O_{19}$. [3]

Wavenumber (cm^{-1})	Modes	Assigned polyhedra
172	E_{1g}	Whole spinel block
181	E_{1g}	Whole spinel block
207	E_{2g}	
283	E_{1g}	Octahedra
318	A_{1g}	Octahedra (mixed)
337	A_{2g}	Octahedra *(12k)*
410	A_{1g}	Octahedra *(12k)* dominated
461	A_{1g}	Octahedra *(2a)*
527	E_{1g}	Octahedra *(2a)*
614	A_{1g}	Octahedra *(4f2)*
684	A_{1g}	Bipyramid *(2b)*
724	A_{1g}	Tetrahedra *(4f1)*

To study further about the structural properties, FTIR spectra of the prepared sample were taken and are shown in Figure 1(c). The band at 441 cm^{-1} is due to the vibration of Fe–O bond in octahedral $4f_2$ site. The bands at 550 and 592 cm^{-1} are due to the vibration of Fe-O bond in the tetrahedral $4f_1$ site [7-9]. The band appeared at 901 cm^{-1} is due to the Ba-O bond stretching [10].

SEM image and surface morphology depicted in Figure 1(d) and Figure 2(a) respectively clearly show platelet shaped grains in the sample. These images also reveal that the hexagonal platelet grains of different sizes are randomly oriented. Moreover, platelets are stacked together in these grains as seen in Figure 1(d). From the EDS spectra shown in the Figure 2(b) it is clear that all the metal cations are present in the sample and the elemental mapping shows uniform distribution of them in the sample as seen in Figure2(c) - Figure2(f).

IOP Conf. Series: Materials Science and Engineering 1233 (2022) 012003 doi:10.1088/1757-899X/1233/1/012003

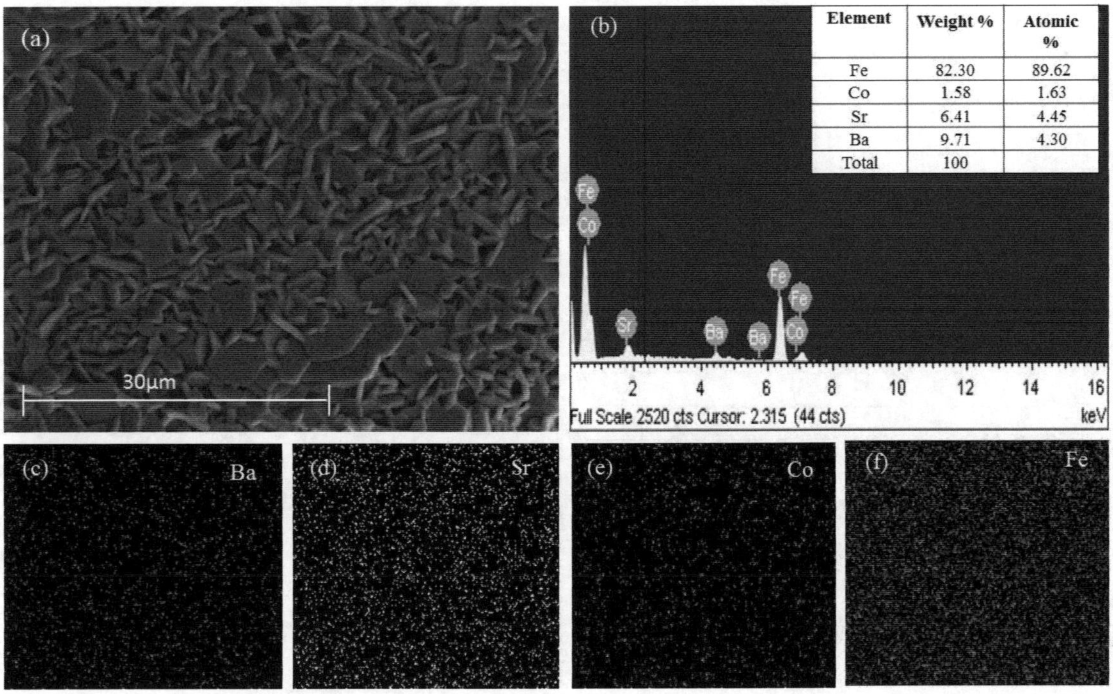

Element	Weight %	Atomic %
Fe	82.30	89.62
Co	1.58	1.63
Sr	6.41	4.45
Ba	9.71	4.30
Total	100	

Figure 2. (a) Surface morphology, (b) EDS spectrum, Elemental mapping of (c) Barium, (d) Strontium, (e) Cobalt and (f) Iron in $Ba_{0.5}Sr_{0.5}Co_{0.2}Fe_{11.8}O_{19}$.

3.2. Magnetic studies

Figure 3(a) shows magnetic hysteresis (M-H) loops of the sample at different temperature. It is seen that the loop is not saturated, even in the highest applied field of 30kOe, at all the measurement temperatures [11]. The spontaneous magnetization (M_S) value was calculated by extrapolating the high field part of M-H loop to the zero applied fields. The M_S value obtained is 62 emu/g at 300K, which is 86% of 72 emu/g observed for $BaFe_{12}O_{19}$ single crystal [12]. The M_S value increases and coercivity (H_C) decreases with the decrease in measurement temperature as seen in Figure 3(b). Similar behaviour has also been reported in $BaFe_{12}O_{19}$ nanoparticles [11]. The highest M_S value of 88 emu/g was observed at 60K. The highest H_C value of 1.22 kOe was observed at 300K and the lowest value of 0.62 kOe was observed at 60K. The observed H_C value of 1.22kOe at 300K is smaller than 5.25kOe observed for $BaFe_{12}O_{19}$ nanoparticles [11].

Ba^{2+} and Sr^{2+} ions have no unpaired electrons. Considering spin contribution only, Fe^{3+} ($3d^5$), Co^{3+} ($3d^6$) and Co^{2+} ($3d^7$) have magnetic moments of $5\mu_B$, $4\mu_B$ and $3\mu_B$ respectively. Out of 24 Fe^{3+} ions in the unit cell of $Ba_{0.5}Sr_{0.5}Fe_{12}O_{19}$, the ions situated at 2a, 2b and 12k sites have up spin and the ions at 4f1 and 4f2 have down spin. If Co^{2+} / Co^{3+} ion replaces Fe^{3+} ion in the up spin state in the unit cell, the net magnetization will be less than that of $Ba_{0.5}Sr_{0.5}Fe_{12}O_{19}$. Moreover, the non-saturation behaviour of the M-H loop may also contribute to observation of lower magnetization in the sample.

Theoretically, single ion anisotropy constants of Fe^{3+} in $BaFe_{12}O_{19}$ were found to be 0.23, 1.40, 0.18, 0.51 and − 0.18 cm^{-1}/ion for 2a, 2b, 4f1, 4f2 and 12 k sites respectively [13]. Among the five different sites, only 12k site makes negative contribution to the anisotropy constant. So, when Co^{2+} / Co^{3+} ion occupy other up spin states, i.e., 2a or 2b, magnetocrystalline anisotropy constant decreases from the high value observed for $BaFe_{12}O_{19}$ [14].

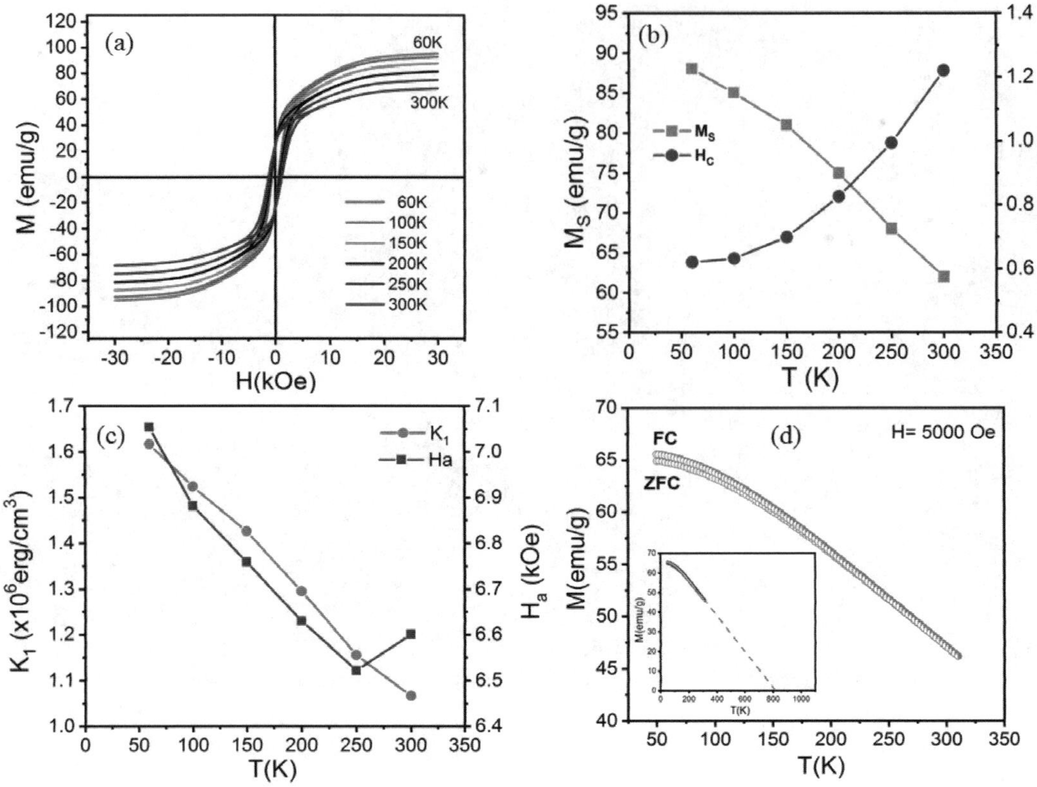

Figure 3. (a) Magnetic hysteresis (M-H) loop, variation of (b) M_S and H_C, (c) K_1 and H_a and (d) ZFC and FC magnetisation with temperature (T) for $Ba_{0.5}Sr_{0.5}Co_{0.2}Fe_{11.8}O_{19}$.

Figure 3(c) shows the variation of magnetocrystalline anisotropy constant (K_1) and anisotropy field (H_a), which were calculated using equations (1) and (2) respectively.

$$K_1 = \frac{M_S}{2}\left(\frac{H_C}{p} + H_b\right) \tag{1}$$

$$H_a = \frac{2K_1}{M_S} \tag{2}$$

where p=0.48 for flake like grains and $H_b = 4\pi M_S$ [12] [15-17]. The observed value of K_1 is 1.07×10^6 erg/cm^3 at 300K and is comparable to the reported value observed for $BaFe_{12}O_{19}$ nanoparticles [17-18] From the Figure 3(c) it is clear that both K_1 and H_a increase with decreasing temperature. Similar behaviour has also been observed in $LaFe_{12}O_{19}$[19].

Zero field cooled (ZFC) and field cooled (FC) magnetization variation with temperature at an applied field of 5kOe is shown the Figure 3(d). It is seen than both ZFC and FC magnetization decrease with temperature. Curie temperature (T_C) was calculated by extrapolating the FC curve to temperature axis (shown as inset in Figure 3(d)) and was found to be 819K compared to the observed T_C of 740K for $BaFe_{12}O_{19}$ and 750K for $SrFe_{12}O_{19}$ [12][20].

4. Conclusion

M-type hexaferrite $Ba_{0.5}Sr_{0.5}Co_{0.2}Fe_{11.8}O_{19}$ was synthesized by ceramic method. Platelet like grains were observed in the sample and elemental distribution analysis confirmed uniform distribution of cations in the sample. From the temperature dependent magnetic studies, magnetisation value of 62 emu/g was observed at 300K, which is 86% of the magnetisation value for M-type Ba-hexaferrite single crystal. Magnetization increased and coercivity decreased with decrease in measurement temperature. Both magnetocrystalline anisotropy constant and anisotropy field increased with decreasing temperature. The observed magnetic behaviour is due to the cation distribution in the lattice sites, nonsaturation of magnetic hysteresis loop and shape anisotropy in the sample.

5. References

[1] Goldman A 2006 *Modern Ferrite Technology* Springer Second Edition.
[2] Pullar R C 2012 Hexagonal ferrites: A review of the synthesis, properties and applications of hexaferrite ceramics *Progress in Materials Science* **57** 1191–1334.
[3] Kreisel J, Lucazeau G and Vincent H 1998 *Journal of Solid State Chemistry* **137** 127–137.
[4] Kumar S, Supriya S and Kar M 2017 *Mater. Res. Express* **4** 126302.
[5] Behera P and Ravi S 2016 *J Supercond Nov Magn* **30** 1453–61.
[6] Chen M S, Shen Z X, Liu X Y and Wang J 1999 *Journal of Materials Research* **15** 483–487.
[7] Zhao W Y, Wei P, Wu X Y, Wang W and Zhang Q J 2008 *J. Appl. Phys.* **103** 063902.
[8] Ram S 1995 *Physical Review B* **51(10)** 6280–86.
[9] Vadivelan S and Jaya N V 2016 *Results in Physics* **6** 843–850.
[10] Pereira F M M, Junior C A R, Santos M R P, Sohn R S T M, Friere F N A, Sasaki J M, de Paiva J A C and Sombra A S B 2008 *J Mater Sci: Mater Electron* **19** 627–638.
[11] Shalini M G, Subha A, Sahu B and Sahoo S C 2019 *Journal of Materials Science: Materials in Electronics* **30** 13647–54.
[12] Shirk B T and Buessem W R 1969 *J. Appl. Phys.* **40** 1294.
[13] Xu Y, Yang G, Chu D, and Zhai H1990 *Phys. Status Solid*. B **157** 685–693.
[14] Li J, Zhang H, Liu Y, Li Q, Zhou T and Yang H 2015 *Applied Physics A* **119** 525–532.
[15] Haneda K, Miyakawa C and Kojima H 1974 *Journal of American Ceramic Society* **57(8)** 354–357.
[16] Mee C D and Jeschke J C 1963 *J. Appl. Phys.* **34(4)** 1271–72.
[17] Krishna murthy J, Mitra C, Ram S and Venimadhav A 2012 *Journal of Alloys and Compounds* **545** 225–230.
[18] Charanjeet S, Narang S B, Hudiara I S, Bai Y and Tabatabaei F 2008 *Materials Research Bulletin* **43** 176–184.
[19] Grössinger R, Küpferling M, Haas M, Müller M, Wiesinger G and Ritter C 2007 *Journal of Magnetism and Magnetic Materials* **310** 2587–89.
[20] Vinnik D A et al 2015 *Materials Chemistry and Physics* **163** 416–420.

Acknowledgments

We thank Centre for Nano Science and Engineering (CeNSE), Indian Institute of Science, Bengaluru, India, for the characterizations of the sample.

IOP Conf. Series: Materials Science and Engineering 1233 (2022) 012004 doi:10.1088/1757-899X/1233/1/012004

Crossover from intrinsic to extrinsic magneto transport in $La_{1-x}Pb_xMnO_3$ perovskite systems

M Sravan Kumar[1,2], P Bala Bhaskar[3], G Narsinga Rao[4,5] and D. Suresh Babu[1]

[1]Department of Physics, Osmania University, Hyderabad 500007, Telangana, India.
[2]St. Joseph's Degree & PG College, King Koti, Hyderabad 500029, India
[3]Government City College (Autonomous), Hyderabad - 500002, Telangana, India
[4]Marri Laxman Reddy Institute of Technology and Management, Hyderabad 500043, Telangana, India
[5]Author to whom any correspondence should be addressed: nsrao@mlritm.ac.in

Abstract. The intrinsic to extrinsic magneto transport of Pb doping and intermediate grinding on transport as well as magnetic properties in $La_{1-x}Pb_xMnO_3$ (x = 0, 0.15, 0.25 and 0.4) has been investigated. Double exchange related electron transport is supposed to prevail in case of highly doped samples whereas spin dependent interfacial tunneling of conduction electrons along with double exchange related electron transport dominate conduction process in case of less Pb substituted samples. It has been shown that with repeated grinding and increase in doping percentage (within the doping limit), it is possible to have relatively narrow grain size distribution, which is a major cause of gradual disappearance of the intergrain interfacial tunneling (extrinsic) resistance leaving only intragrain (intrinsic) transport properties in the high temperature region. At low temperature the existence of spin-glass insulator behaviour were observed, which get stronger with decrease in Pb content. Thus with decrease in Pb doping, it has been possible to generate relatively wide distribution of grain size with probable increase in misorientation angle in the grain boundary region. This reveal a crossover from major intrinsic to combination of extrinsic, intrinsic magnetotransport and spin-glass behaviour.

1. Introduction

Recent upsurge in research for mixed valence perovskite manganites is mainly due to their large negative magnetoresistance (MR), termed as colossal magnetoresistance, behavior near Curie temperature [1], T_c, which readily find the potential application in information storage industry as magneto resistive read heads, providing greater storage density than ordinary inductive devices [2]. Prevailing class of mixed valence perovskite manganites are of general formula $R_{1-x}A_xMnO_3$ where R- rare earth element and A- divalent alkaline-earth metal [3]. Most of these materials are reported to undergo a transition from insulator to metallic phase at a temperature almost coincident with paramagnetic to ferromagnetic phase transition temperature (T_c). The ground state of stoichiometric parent compound $LaMnO_3$ is an insulating material, which orders antiferromagnetically [4] due to orbital ordering and super exchange interactions among different Mn^{3+} layers, though within a layer, Mn^{3+} ions are coupled ferromagnetically [5]. In spite of enormous theoretical and experimental activity in the area of phase separation in these manganites, still many questions are remaining open. Intrinsic effects are believed to arise for thermodynamic reasons such as structural and magnetic inhomogeneity in manganites. Extrinsic inhomogeneity arises due to various technological factors in the sample preparation. They can cause inhomogeneity in the chemical composition (oxygen content), structural inhomogeneity (polycrystalline or even granular structure), strain inhomogeneity, and so on.

Partial doping of trivalent cation with divalent one creates holes (Mn^{4+}) giving rise to itinerant eg electrons hopping between spin aligned (due to strong on site Hund coupling) Mn^{3+} (t_{2g}^{3} e_{g}^{1}) and Mn^{4+} (t_{2g}^{3}) ions through oxygen ion producing metallic conductivity [6] and ferromagnetism. This phenomenon is known as Double Exchange mechanism [7]. In case of single crystals of doped manganites, MR (termed as intrinsic) is maxima near T_c in relation to reduction of spin fluctuation by the application of magnetic field and becomes almost zero at low temperature when the magnetization gets saturated. A very high field, of the order of few Tesla, is required to get the MR effect in single crystals making it poorly sensitive at the low field region. For polycrystalline samples the situation is completely different, where grain boundaries/interfaces play a major role in determining the overall magnetotransport properties. A negative MR (termed as extrinsic) with a steep rise at very low field and low temperature is generally observed in polycrystalline perovskite manganites as an effect of either spin polarized tunneling between neighboring highly spin polarized grains (non percolative configuration) [8] or spin dependent scattering at the interfaces/grain (percolative configuration) [9] boundaries having canted magnetic structure. The high field MR, which has been explained by the spin flip of tunneling electrons at the barrier in case of spin dependent tunneling mechanism and by the progressive saturation of the grain boundary magnetization in case of spin dependent scattering one, remain comparatively temperature independent in both the configurations. Recently, the manganites having both the signature of intrinsic (intragrain) as well as extrinsic (intergrain tunneling) magnetoresistance at relatively higher temperatures have been reported [10-12], which have never been seen in other granular ferromagnets [13]. This has been explained as due to the magnetic inhomogeneity vis-à-vis double exchange interactions between the surface/interface and the cores.

To delineate the contribution of the grain boundaries/interfaces in the overall magneto transport properties, in this paper, the effect of Pb doping in La site of $LaMnO_3$ and preparation conditions are investigated.

2. Experimental details:

$La_{1-x}Pb_xMnO_3$ (x = 0, 0.15, 0.25 and 0.4) polycrystalline samples have been prepared by standard solid state reaction method with high purity powders of La_2O_3, PbO and MnO_2 (99.9% purity) taken in proper stoichiometric ratio. Powders are ground in a mortar in acetone medium for homogeneous mixing and calcined at 1173 K for 24 hours in air. The calcined powders are ground and then cold-pressed into pellets with 5000 lb pressure for 5 minutes and sintered at 1473 K for 24 hours in air, which we refer as first grinding samples. The pellets are again reground and cold-pressed into pellets with 5000 lb pressure for 5 minutes and sintered at 1473 K for 24 hours in air, which we refer as second grinding samples. The formation of perovskite single phase has been confirmed by powder X-ray diffraction pattern using Philips PW-1817 with Cu-Kα radiation. The grain size has been estimated from the Scanning Electron Micrographs using Hitachi S-3000. The DC resistance and magnetoresistance have been measured by standard four probe technique using close cycle helium cryostat from 10 K to 300 K placed in a conventional water cooled electro magnet with a field from – 8.8 KOe to + 8.8 KOe. The dc field cooled (FC) and zero field cooled (ZFC) magnetizations have been measured in commercial squid magnetometer from 2 K to 350 K with a field H = 100 Oe.

3. Results and Discussions:

Figure 1 shows powder X-ray diffraction patterns of polycrystalline $La_{1-x}Pb_xMnO_3$ (x = 0, 0.15, 0.25 and 0.4) samples after first and second grinding samples. From these diffraction patterns, it is evident that there is no extra Lead or any other oxide impurity peak, confirming the formation of single phase in all the samples of first and second grinding. This necessarily means, though the ionic radius of Pb is slightly higher than that of La, it is very likely, Pb has properly been substituted in La site. We have calculated the corresponding d_{hkl} values and lattice parameters for each sample, which very closely relate to the rhombohedral structure [14] with a space group $R\bar{3}c$. The ionic radius of Pb being bigger than that of La, it seems, the lattice should have dilated in case of proper substitution but since the Pb is in its divalent state and La being in its trivalent one, the Pb substituted samples should have less open structure involving less oxygen and effectively decreased lattice parameters on the other hand. That is why in our samples lattice parameters remain almost same throughout the doping limit.

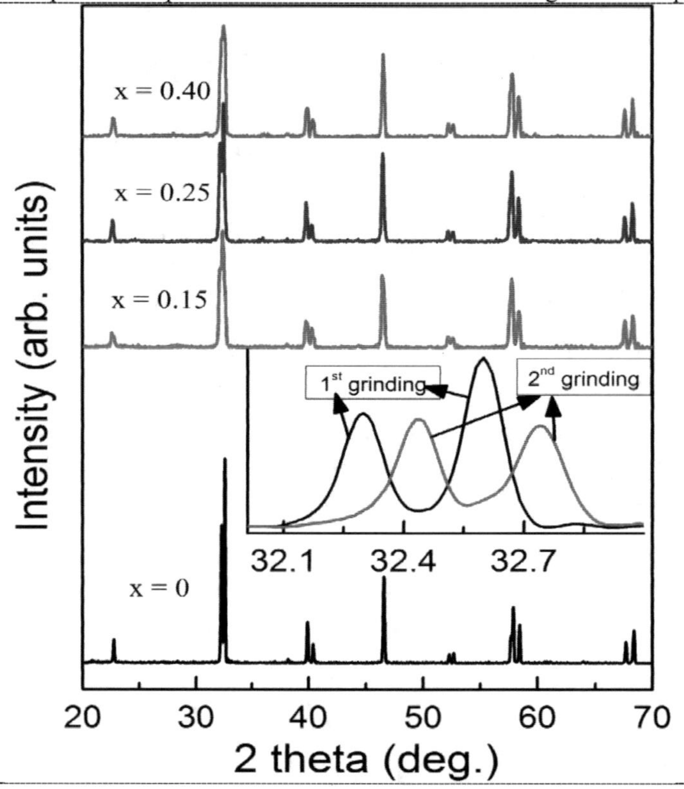

Figure 1. Powder X-ray diffraction patterns of polycrystalline $La_{1-x}Pb_xMnO_3$ for x = 0, 0.15, 0.25 and x = 0.4 samples after first grinding. In inset the change in first and second grinding samples.

Figure 2 shows the typical scanning electron micrographs for x = 0.25 after (a) first and (b) second grinding, which are the typical micrographs for all other samples after respective grindings. From the micrographs it is evident that increase of Pb ion increases the average grain size and decreases the grain size distribution. In principle it is quite possible because addition of Pb acts as nucleation center for growing perovskite crystalline phase, which favors grain growth compared to the pristine oxide at a particular sintering temperature. The number of such nucleation centers increases with increase in Pb content, effectively increasing the grain size and decreasing the grain size distribution. It is also evident that with repeated grinding for a particular composition, the sample becomes more homogeneous with a grain growth and relatively narrow particle size distribution. These have been correlated further with transport/magnetotransport properties in subsequent part of our discussion.

Figure 2. Scanning electron micrographs for x = 0 sample (a) after first grinding (b) after second grinding.

Figure 3 shows the temperature variation of resistivity ρ(T) in zero field with different Pb content samples after first and second grindings. It is found that as we increase the Pb content the resistivity decreases. The ρ(T) curve of x = 0 sample shows continues increase with decreasing temperature, as an insulating behavior. The ρ(T) curves shows sharp metal insulator resistivity peak (T_{P1}) at 335 K, 285 K, and 255 K for the x = 0.4, 0.25 and 0.15 samples, respectively along with broad maxima (T_{P2}) in the resistivity curves at successively decreasing temperatures. This broad maxima shift to lower temperatures with decrease of Lead content in the La site. The decrease in overall resistivity and shifting of first resistivity peak (T_{P1}) to the higher temperature side with increase in Pb content vis-à-vis Mn^{4+} concentration are consistent with the results obtained from other divalent doped manganites, which signify that long range ferromagnetic order have been formed at successively higher temperatures with increase of Pb content. The shift of the second resistivity maxima (T_{P2}) to the higher temperature side as well as diminishing peak height with increase in Pb content can be correlated with the increase of individual grain size and decrease of grain size distribution with an effective decrease of surface/interface area, i.e. interfacial scattering of conduction electrons. All these samples show an upturn in resistivity (T_{upturn}) at low temperature. The temperature of upturn in resistivity increases with Pb doping it goes to further lower temperature side.

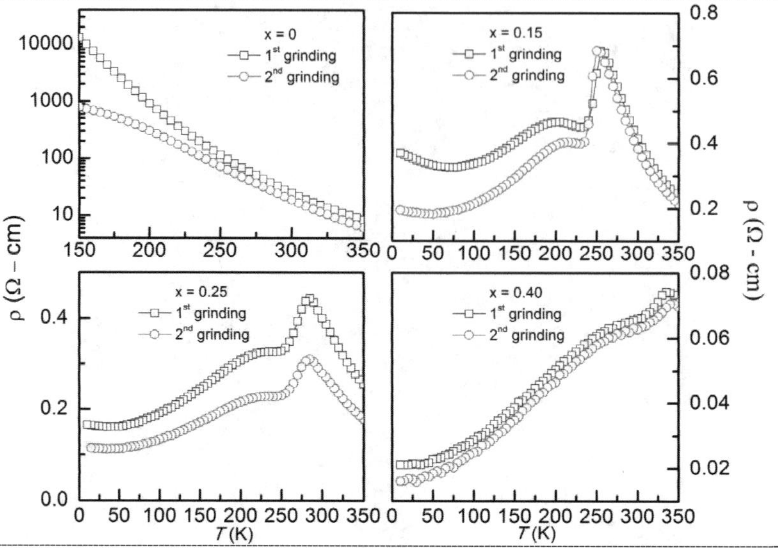

Figure 3. The variation of resistivity in zero field with respect to temperature and Pb content for samples after first grinding and second grinding samples.

Figure 4 show the variation of MR ratio with respect to temperature for the different composition after first grinding. From the figures it is evident that nature of different resistivity maxima at H = 0 for a particular Lead doped sample is different, signifying the difference in underlying transport / magnetotransport mechanisms. This can also be substantiated from the resistivity curves under application of magnetic field. In the case of first resistivity peak (T_{P1}), the peak shifts to higher temperature with field, which is characteristic feature of double exchange mechanism18. Whereas in case of second resistivity peak (T_{P2}), the temperature of resistivity maxima remain constant even with application of magnetic field signifying the possibility of some other mechanism i.e. appearance of interfacial tunneling resistance and magnetoresistance between different grains. These have been further elaborated in the latter part of our discussion.

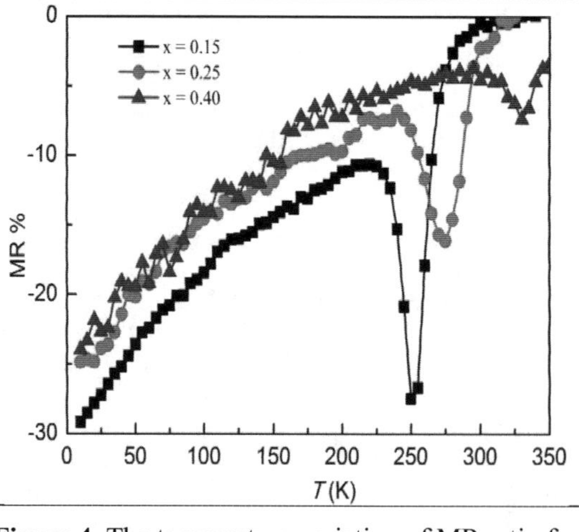

Figure 4. The temperature variation of MR ratio for the Pb doped samples after first grinding.

It is also evident from figure 3 that with repeated grinding the temperature of first resistivity peak (T_{P1}) remains almost same with relatively sharper transition width, whereas the temperature of second resistivity peak (T_{P2}) shifts to much higher temperature with diminished peak height compared to first grinding sample, effectively producing very high MR, which is intrinsic in nature, evident from the figure 4. After second grinding the entire composition becomes more homogeneous with decreasing thickness of paramagnetic/antiferromagnetic interfaces (evident from the micrographs and can be substantiated by the narrow transition width), which facilitates the establishment of long range ferromagnetic ordering at relatively higher temperature vis-à-vis sharp fall in resistivity immediately after transition in one hand as well as diminished second resistivity peak height due to reduction in spin dependent interfacial tunneling resistance on the other. We presume high MR ratio is the manifestation of simultaneous effect of these two.

In hole doped perovskite manganites adjacent Mn^{3+} (t_{2g}^{3} e_g^{1} ; S = 2) and Mn^{4+} (t_{2g}^{3} e_g^{0}; S = 3/2) ions are ferromagnetically coupled with each other through oxygen ion producing Zener pairs [15-16], where delocalized eg electrons due to their intrinsic spin and strong correlation with localized t_{2g}^{3} electrons (Hund's coupling) hop from a Mn^{3+} site to a Mn^{4+} site through oxygen ion, having the same spin orientation. These Zener pair's segregates into short range ferromagnetically ordered domains due to dynamic phase segregation in the manganese based perovskite oxides, which remain isolated at relatively higher temperature causing the system to be paramagnetic as a whole. With decreasing temperature, size of the each ferromagnetically ordered domain increases at the expense of paramagnetic region producing fm-pm-fm network of magnetic coupled junctions. Since in the ferromagnetic region all the spins are aligned in the same direction, but Zener pairs move easily giving rise to metallicity. Whereas in the paramagnetic region, the movement of the same get hindered due to the disorder scattering from charge carriers with different spin orientations giving rise to insulating/semiconducting behavior. On further cooling, the thickness of the paramagnetic junction gets reduced and two neighbouring ferromagnetic regions overlap with each other to produce a long range ferromagnetic order giving rise to metallic conductivity for the system as a whole, which shows metal–insulator transition in the resistivity curve.

The characteristically different broad second resistivity maxima (T_{P2}) for all the compositions are supposed to originate from different mechanism than the pure double exchange mechanism responsible for the sharp resistivity peaks in the samples, which has been described as follows: Incorporation of Pb ion increases the average grain size and decreases the grain size distribution, which in turn decreases the total surface/interface area. In granular perovskites, the magnetic structure of the interface between two neighbouring grains acting as potential barrier plays a major role in determining the overall magneto-transport properties. These interfaces being amorphous in nature will have large number of dangling bonds and hence will have quite different (random) magnetic structure, effectively producing much weaker coupling, i.e. double exchange interactions among the present magnetic ions with respect to that of cores providing intergrain potential barrier. Recently it has been found in $La_{0.7}Sr_{0.3}MnO_3$ using spin resolved photoemission spectroscopy [17] with a 5Å probing depth that the T_C is quite depressed at the surface/interface than that of the core, proving further, magnetic disorder at the interface. It has been theoretically shown that the broad resistivity maxima can appear from double exchange related spin dependent interfacial tunneling due to the difference in magnetic order between the surface/interface and core [10,18]. It is evident from figure 4 that there is no MR % peak at around the resistivity maxima (T_{P2}) in lead doped samples, which indeed confirms the appearance of resistivity maxima due to extrinsic type spin dependent interfacial tunneling arising from the magnetic inhomogeneity, i.e. difference in magnetic order between surface/interface and core.

In one of the recent references [19] Pb content has been varied from $0.1 < x < 0.4$ in $La_{1-x}Pb_xMnO_3$ system prepared by repeated grinding, pelletizing and prolonged annealing at 950^0C without any signature of second resistivity maxima in the corresponding resistivity curves. We feel that is because, as in our second grinding sample, repeated grinding made samples more homogeneous with narrow grain size distribution and subsequently larger grains developed due to prolonged heat treatment with less volume percentage of interfaces, giving only double exchange related metal to insulator transition near Curie temperature.

Hence, two MR regimes can be expected. Initially a regime corresponding technical magnetization where the alignment in the neighbouring domains is obtained under application of low magnetic field and then the regime corresponding to saturation of the grain boundary magnetization under a relatively high field. The magnetoresistance will fall very sharply at low magnetic field and will decrease very slowly with high magnetic field for spin dependent interfacial tunnelling [17] whereas the same will decrease almost linearly in the entire field region in case of intrinsic magnetotransport. In the light of above discussion, from figures 4 it can be clearly understood that the MR ratios in all samples are due to spin dependent interfacial tunnelling both at 10 K and at second resistivity peak temperatures (T_{P2}) whereas the MR ratios in case of first resistivity peak temperatures (T_{P1}) is due to intrinsic magnetotransport. With decrease in Pb doping it has been possible to show crossover from major intrinsic to combination of extrinsic and intrinsic magnetotransport behavior with high MR ratio at relatively low field. However, repeated grinding for a particular composition changes the ratio of intrinsic to extrinsic transport giving rise to further higher MR (intrinsic in nature) ratio, in $La_{1-x}Pb_xMnO_3$ perovskite systems.

4. Conclusions:

The intrinsic to extrinsic magneto transport of Pb doping and intermediate grinding on transport as well as magnetic properties in $La_{1-x}Pb_xMnO_3$ (x = 0, 0.15, 0.25 and 0.4) has been investigated. Double exchange related electron transport is supposed to prevail in case of highly doped samples whereas spin dependent interfacial tunneling of conduction electrons along with double exchange related electron transport dominate conduction process in case of less Pb substituted samples. It has been shown that with repeated grinding as well as increase in doping percentage (within the doping limit), it is possible to have relatively narrow grain size distribution, which is a major cause of gradual disappearance of the intergrain interfacial tunneling (extrinsic) resistance leaving only intragrain (intrinsic) transport properties in the high temperature region. At low temperature the existence of spin-glass insulator behaviors, which gets stronger with decrease in Pb content, have been obtained in all samples. Thus with decrease in Pb doping, it has been possible to generate relatively wide distribution of grain size with probable increase in misorientation angle in the grain boundary region, to show a crossover from major intrinsic to combination of extrinsic, intrinsic magneto transport and spin-glass behavior, which produces high MR ratio (~ 35 %) upon grinding at relatively low field and high temperature in $La_{1-x}Pb_xMnO_3$ perovskite systems.

Acknowledgement:
MSK would like to thank Rev.Fr. Dr. D. Sunder Reddy, Principal, St. Josephs Degree & PG College (Autonomous) for his motivation and encouragement.

References:

[1] Jin S, Tiefel T H, McCormack M, Fastnacht R A, Ramesh R, Chen H L 1994 Science, 264, 413
[2] White R L 1992 IEEE Trans. Magn. 28 2482
[3] Imada M, Fujimori A and Tokura Y 1998 Rev. Mod. Phys. 70 1039
[4] Wollan E O and Koehler W C 1955 Phys. Rev. 100 545
[5] Zaanen J, Sawatzky G A and Allen J W 1985 Phys. Rev. lett. 55 418
[6] A. Krichene A, Solanki P, Royaprol S, Ganeshan V, Houjelben W and Kuberkar D G 2015 Ceramics International 41 2637
[7] Zener C 1951 Phys. Rev. 81 440; 82, 403
[8] Hwang H Y, Cheong S W, Ong N P and Batlogg B 1996 Phys. Rev. Lett. 77 2041
[9] Isaac S P, Mathur N D, Evetts J E and Blamire M G 1998 Appl. Phys. Lett., 72 2038
[10] Zhang N, Ding W, Zhong W, Xing D and Du Y 1997 Phys. Rev. B, 56 8138
[11] Zhang N, Zhong W and Ding W 1999 J. Vac. Sci. Technol. B 17 1050
[12] Wang X L, Gehringer P, Lang W, Horvat J, Liu H K and Dou S X 2001 Solid State Comm. 117 53
[13] Berkowitz A E, Mitchell J R, Carey M J, Young A P, Zhang S, Spada F E, Parker F T, Hutten A and Thomas G 1992 Phys. Rev. Lett. 68 3745
[14] Arulraj A, Mahesh R, Subbanna G N, Mahendiran R, Raychaudhuri A K and Rao C N R 1996 J. Solid State Chem. 127 87
[15] Deshmukh A V, Patil S I, Bhagat S M, Sagdeo P R, Choudhary R J and Phase D M 2009 J. Phys. D: Appl. Phys. 42 185410
[16] Laiho R, Lahderanta E, Salminen J, Lisunov K G and Zakhvalinskii V S 2001 Phys. Rev. B 63 94405
[17] Park J H, Vescovo E, Kim H J, Kwon C, Ramesh R and Venkatesan T 1998 Phys. Rev. Lett. 81 1953
[18] Joonghoe Dho, Kim W S and Hur N H 2002 Phys. Rev. Lett. 89 027202
[19] Powell A V, Herwig C, Colgan D C and A Gavin Whittaker 1998 J. Mater. Chem. 8 119

IOP Conf. Series: Materials Science and Engineering 1233 (2022) 012005 doi:10.1088/1757-899X/1233/1/012005

Study of the utility of Fe_2O_3-Au nanocomposite in photodynamic therapy

Divya Rajan[1,2,*], Frincy Francis[1], Daniya Thomas[1], Meenakshi Ramankutty[1] and Santhi Ani Joseph [1]

[1]Department of Physics and Centre for Research, St.Teresa's College (Autonomous), Ernakulam,
PIN 682011, India
[2]Department of Physics, St.Peter's College, Kolenchery,
PIN 682311, India
*e-mail: divyarajank@gmail.com

Abstract. A novel multifunctional nanofluid composite of Iron oxide/Gold (**Fe_2O_3-Au**) was prepared using simple wet chemical method by reducing Chloroauric acid using trisodium citrate in the presence of Fe_2O_3 stock solution. The optical and morphological characterization confirmed the formation of hybrid structure of Iron Oxide with Gold nanoparticles. In-vitro cytoprotective studies of the sample was carried out in L929 cells. The study showed an increase up to 15% in the cell viability by the co-administration of 25 μL of Fe_2O_3-Au nanofluid in the L929 cells, which suffered significant reduction in cell viability due to laser irradiation. Thus, the sample is found to possess potential application in Photodynamic therapy.

1.Introduction

Composite nanoparticles have attracted great attention in the present times because of the combined physio-chemical properties of multi-materials and a wide range of potential applications. Gold and iron oxide are two promising candidates to form a multifunctional nanocomposite whose properties opens the windows for new applications if combined together. Nanogold possesses unique characteristics because of the phenomenon of localized surface plasmon resonance (SPR). Gold nanoparticles exhibits biocompatibility, non-cytotoxicity and optoelectronic properties which has been utilized in biological applications like biosensing, and in optical applications [1-3]. As gold is an expensive material, it is feasible to combine gold with a material which is less expensive and still has potential applications. Iron oxide is considered to be a very important semiconducting materials as it is economical, nontoxic, biocompatible and environmentally friendly [4]. Therefore, it is trustworthy to focus on the fabrication of the gold/iron oxide nanocomposites so as to develop multifunctional materials that possess the properties of both gold and iron oxide.

Photodynamic therapy[PDT] is a minimally invasive technique that can be curative as well as preventive in the case of many diseases , especially in the treatment of cancer.[5]Nanoparticles finds applications in photodynamic therapy as it can overcome many of the limitations of the traditional techniques.[6] One advantage of nanoparticles in PDT is the efficiency to increase the amount of photosensitizer to be delivered to the target cells due to the large surface to volume ratio.[7]It can also ensure the timely release of the photosensitizer and thus can prevent the nonspecific accumulation in normal tissues.

Another advantage is that their surface can be modified by targeting agents to alter its properties and can improve the abilities.[8]. In this work we are trying to explore the functionality of Fe_2O_3-Au nanocomposite solution in photodynamic therapy.

2. Sample Preparation

The method adopted for the synthesis of hybrid Fe_2O_3-Au nanoparticles is a simple, wet chemical method which involves a chemical reduction process using a reducing agent. The chemicals used were of analytical grade and were not further purified. The precursors used for the synthesis are Iron (III) oxide (Fe_2O_3), Chloroauric acid ($HAuCl_4$) used to form neutral gold, Sodium citrate ($Na_3C_6H_5O_7.2H_2O$) which act as a reducing agent to form Au (0), all purchased from Merck. Stock solutions of 25mM Fe_2O_3, 0.1 M sodium citrate and 0.01 M chloroauric acid were prepared in deionized water. To 10ml deionized water in a beaker, 100μl of Fe_2O_3 stock solution was added and the solution was heated while stirring for approximately 5 minutes. Next, 1ml of 0.1 M sodium citrate solution was added and the solution was brought to boil. Then 250 μl of 0.01 M chloroauric acid was added to the boiling solution and continued heating at 100°C for 10 minutes. After 2-3 minutes the solution was seen to turn reddish brown indicating the formation of gold nanoparticles. The solution was removed from the heating block and allowed to cool off at room temperature for 1-2 hours. The sample was purified by centrifugation for 7 minutes at 8000 rpm. Later on, the supernatant was removed and the sample was redispersed in distilled water up to 10ml. [9].

Then the optical, morphological and elemental characterization and further the Bioimaging, Cellular uptake and Cytoprotective studies were conducted.

3. Characterization

The measurement of the sample's light absorption at room temperature is carried out using a JASCO spectrophotometer with a resolution of 1.0nm and wavelength range 200 to 900 nm. The fluorescence emission from the same is recorded using Flouoromax 4 spectrofluorometer. The synthesized Fe_2O_3-Au sample was characterized by FESEM using Carl Zeiss Zigma TM instrument with specifications of 5kVand 10X - 1,000,000X equipped with an energy dispersive X-ray spectroscopy (EDS) to investigate morphology and elemental composition.

3.1. Optical Characterization

The UV-Vis absorption spectrum of the sample is shown in Figure 1. In the spectrum, a broad absorbance peak is observed between 350–400 nm which is due to the $\mathbf{Fe^{3+}}$ ligand to metal charge transfer transitions and a shoulder peak between 600 and 650 nm which is due to the pair excitation by the simultaneous excitation of magnetically coupled adjacent $\mathbf{Fe^{3+}}$ cations in Fe_2O_3 [10]. The peak at 526 nm corresponds to the plasmonic peak of gold nanoparticles.

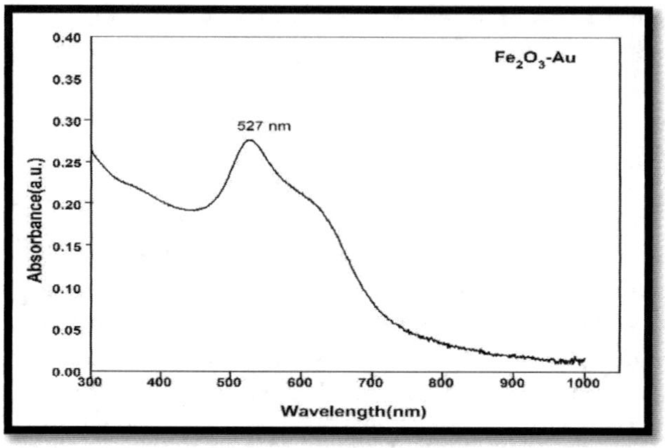

Figure 1. UV-Visible absorption spectrum of Fe_2O_3-Au nanocomposite

The PL emission spectrum of Fe_2O_3-Au nanofluid was measured with an excitation wavelength of 450 nm as shown in Figure 2. When excited with 450 nm, a strong peak appeared at 532 nm which may correspond to the plasmon band of the noble metal. The PL emission of Au nanoparticles can be attributed to the inter-band transitions of d band electrons into the conduction band (sp band) and subsequent radiative recombination with holes [11].

Photoluminescence is not exhibited by α-Fe_2O_3 in the bulk, because of localized d–d transitions and resonant energy transfer between cations, but it can be observed when the size of Fe_2O_3 is reduced to the nanometer scale.[12] When excited with a lower wavelength of 290 nm, we got a sharp peak at 322nm as in Figure 3. This luminescence may originate from electronic transition within the iron oxide (α-Fe_2O_3) nanoparticle. The lowest exciton state of Fe_2O_3 arises from excitation between the mixture of Fe (3d) and O (2p) states in the valance band, and the Conduction Band which derives from Fe (4s) states. Under excitation of 290 nm, an electron is excited from the valance band to the conduction band, leaving a hole in the valance band. The exciton (-) and the hole (+) move freely and forms an exciton, the recombination of the electron and hole in the exciton yields the photoluminescence emission. A broad emission is observed around 400 nm which may relate to deep level emissions and mainly result from defects such as interstitials and oxygen vacancies. [13].

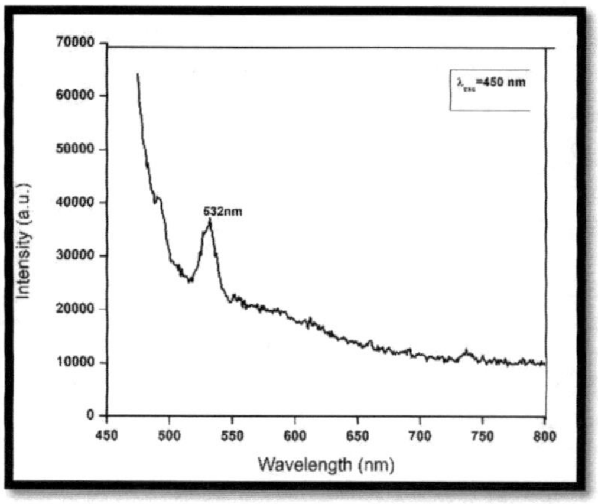

Figure 2.PL spectrum of Fe_2O_3-Au nanocomposite at excitation wavelength of 450 nm

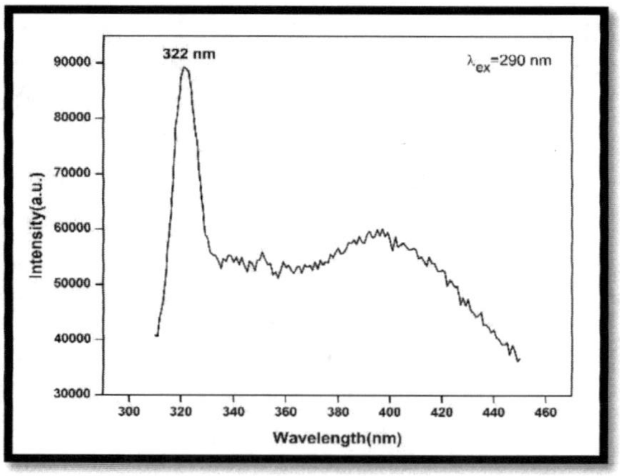

Figure 3.PL spectrum of Fe_2O_3-Au nanocomposite at excitation wavelength of 290 nm

3.2. Morphological /Elemental Characterization

FESEM image confirmed the formation of hybrid structure of Fe_2O_3-Au where small bright gold nanoparticles are decorated on the surface of irregular shaped Fe_2O_3 nanoparticles .(Figure 4)

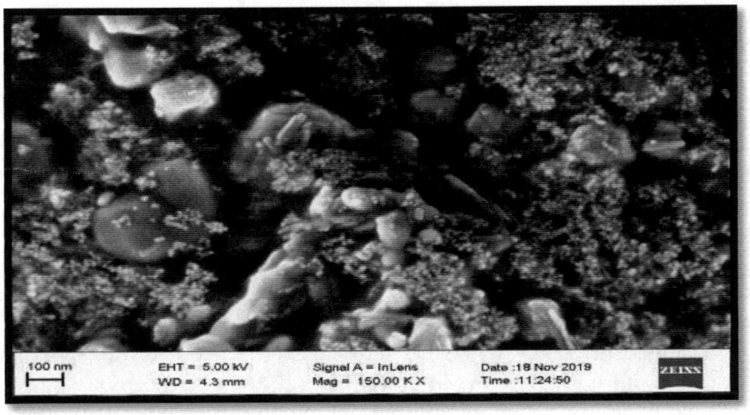

Figure 4. FESEM image of Fe_2O_3-Au nanocomposite

The EDS spectra of Fe_2O_3-Au nanocomposite confirmed the presence of elements Fe, O and Au in the prepared sample (Figure 5). The EDS data of the sample is provided in Table 1.

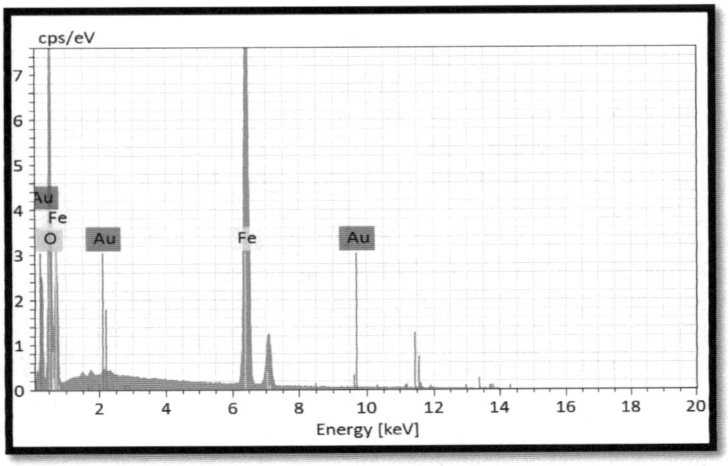

Figure 5. EDS spectrum of Fe_2O_3-Au nanocomposite

Table 1. *EDS data of Fe_2O_3-Au nanocomposite*

Element	Atomic number	*Mass %*	Atomic %
Oxygen	8	28.91	64.41
Iron	26	55.64	35.52
Gold	79	0.41	0.07

4. Biomedical application study of Fe2O3- Au nano composite in Photodynamic Therapy

Photodynamic therapy (PDT) is a relatively new and non-invasive therapeutic technique. It is used in the treatment of non-cancerous diseases as well as different forms of cancers. It has reliable results and the advantage of co-administration possibility with other therapeutic protocols. [14,15] PDT also finds application in areas like dermatology, gynecology and urology. It is based on the local or systemic application of a photosensitizer, which absorb light of the appropriate wavelength, initiating the activation processes leading to the selective destruction of the inappropriate cells. The selective action makes it suitable for a vast majority of patients. [16,17] In our study, we investigated the suitability of our sample Fe_2O_3-Au which is bio safe for application in photodynamic therapy.

4.1. In vitro Cytoprotective effect determination by MTT Assay for L929 Cells using Fe_2O_3-Au nanocomposite

L929 (Mouse Fibroblast) cell line were initially procured from National Centre for Cell Sciences (NCCS), Pune, India and maintained in Dulbecco's Modified Eagles medium (DMEM) (Sigma Aldrich, USA). The cell line was cultured in 25 cm^2 tissue culture flask with DMEM supplemented with 10% FBS, L-glutamine, sodium bicarbonate and antibiotic solution containing: Penicillin (100U/ml), Streptomycin (100µg/ml), and Amphotericin B (2.5µg/ml). Cultured cell lines were kept at 37°C in a humidified 5% CO_2 incubator (NBS Eppendorf, Germany). The viability of cells was evaluated by direct observation of cells by Inverted phase contrast microscope followed by MTT assay method.

Cells seeding in 96 well plate:
Two days old confluent monolayer of cells were trypsinized and the cells were suspended in 10% growth medium, 100µl cell suspension (5x10^4 cells/well) was seeded in 96 well tissue culture plate and incubated at 37°C in a humidified 5% CO_2 incubator.

Preparation of compound stock:
1mg of the sample was weighed and completely dissolved in 1mL DMEM using a cyclomixer. The extract solution was filtered through 0.22 µm Millipore syringe filter to ensure the sterility.

Cytotoxicity Evaluation:
After attaining sufficient growth, the cells were exposed with laser for one hour. The medium was removed and the prepared extracts in 5% DMEM were five times serially diluted by two-fold dilution (100µL, 50µL, 25µL, 12.5µL, 6.25µL in 500µl of 5% DMEM) and each concentration of 100µl were added in triplicates to the respective wells and incubated at 37°C in a humidified 5% CO_2 incubator for 24 hours. Non-treated control cells and cells exposed with laser were also maintained.

Cytotoxicity Assay by Direct Microscopic observation:
Entire plate was observed after 24 hours of incubation in an inverted phase contrast tissue culture microscope (Olympus CKX41 with Optika Pro5 CCD camera) and microscopic observation were recorded as images. Any detectable changes in the morphology of the cells, such as rounding or shrinking of cells, granulation and vacuolization in the cytoplasm of the cells were considered as indicators of cytotoxicity.

Cytotoxicity Assay by MTT Method:
Fifteen mg of MTT (Sigma, M-5655) was reconstituted in 3 ml PBS until completely dissolved and sterilized by filter sterilization. After 24 hours of incubation period, the sample content in wells were removed and 30µl of reconstituted MTT solution was added to all test and cell control wells, the plate was gently shaken well, then incubated at 37°C in a humidified 5% CO_2 incubator for 4 hours. After the incubation period, the supernatant was removed and 100µl of MTT Solubilization Solution (Dimethyl

Sulphoxide, DMSO Sigma Aldrich, USA) was added and the wells were mixed gently by pipetting up and down in order to solubilize the formazan crystals. The absorbance values were measured by using microplate reader at a wavelength of 540 nm [18]

5. Results and discussions

The percentage of growth inhibition was calculated using the formula:

$$\% \, of \, viability = \frac{Mean \, OD(optical \, density\,)samples \, \times 100}{Mean \, OD \, of \, control \, group}$$

Table 2. *The percentage viability of L929 cells as control and under the influence of $Fe_2\,O_3$ –Au nanoparticles in varying concentration.*

Concentration(µg/mL)	OD I	OD II	OD III	Average OD	Percentage Viability
Control	0.4582	0.4936	0.4652	0.4723	100
Laser exposed cells	0.3409	0.3470	0.3211	0.3363	71.21
Sample					
6.25	0.3746	0.3573	0.3710	0.3676	77.84
12.5	0.3777	0.3875	0.3733	0.3795	80.35
25	0.4276	0.4129	0.3833	0.4079	86.37
50	0.4115	0.3788	0.3800	0.3901	82.60
100	0.3820	0.3522	0.3741	0.3694	78.22

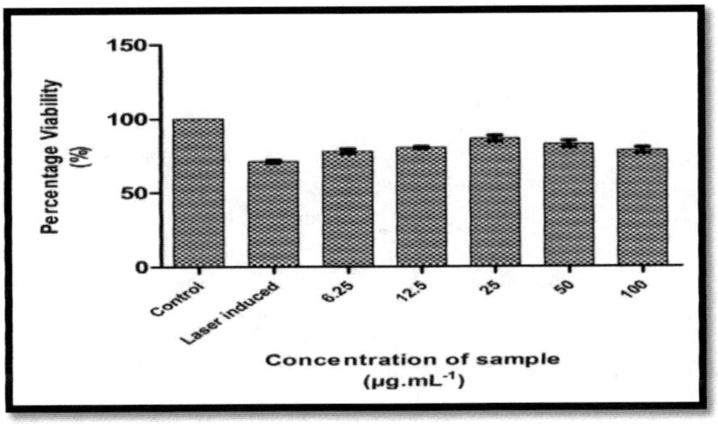

Figure 6. Graphical representation of in vitro cytoprotective effect of Fe2O3-Au on laser irradiated L929 cells

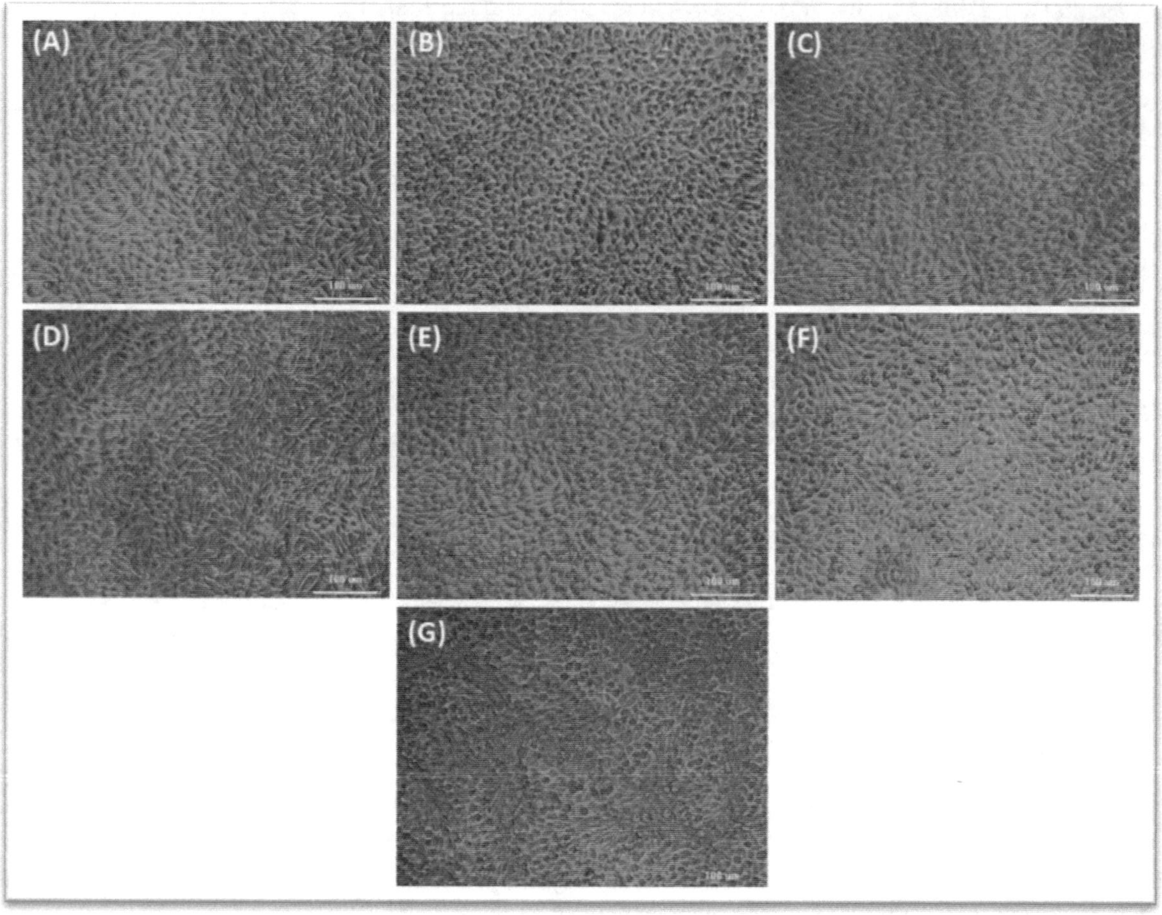

Figure 7: Determination of *in vitro* cytoprotective effect of Fe_2O_3-Au on laser irradiated L929 cells **(A)** untreated L929 cells and laser irradiated cells treated with Fe_2O_3-Au nano composite **(B)** laser irradiated L929 cells **(C)** 6.25 **(D)** 12.5 **(E)** 25 **(F)** 50 **(G)** 100 µL (Magnification 20X).

The study for In-vitro cytoprotective effect determination by MTT Assay for L929 cells using Fe_2O_3-Au nanocomposite reveals that the Laser exposure (532nm) to L929 cells has induced toxicity and reduced the cell viability to 71.21% suggesting cell death. The percentage viability of L929 cells as control and under the influence of $Fe_2 O_3$-Au nanoparticles in varying concentration (Table 2) shows that the compound has afforded protective effects at lower concentration and a maximum increase of viability to 86.37 % was observed in 25 ug/ml concentration of Fe_2O_3-Au. Further increase in concentration was found to reduce the viability which suggest an optimum value of Fe2O3-Au nanocomposite sample for the In-vitro cytoprotective studies. (Figure 7)

6. Conclusions

In the present work, we have synthesized Fe_2O_3-Au nanocomposite using simple wet chemical method. The linear optical characterization including absorption and photoluminescence confirmed the formation of Fe_2O_3-Au nanocomposite. The morphological characterisation using FESEM revealed the formation of decorated gold nanoparticles on the surface of irregular shaped Fe_2O_3. The EDS spectra of Fe_2O_3-Au nanocomposite confirmed the presence of elements Fe, O and Au in the prepared sample. From the studies for Invitro Cytoprotective effect determination by MTT Assay for L929 cells, it is clear that there is a significant decrease of cell viability to 71.21 % upon treatment with laser for 1 hour which was increased to 86.37 % by 25 μL of Fe2O3-Au nanocomposite. There was concomitant reversal of morphological characteristics as determined by phase contrast microscopy. Laser irradiation was found to induce cytotoxicity of fibroblast cells as determined by MTT assay and co administration of Fe2O3-Au nanocomposite increases the viability of cells up to 15 % at a concentration of 25 μL of nanocomposite, thus proved to be a suitable candidate for photodynamic therapy. It has been reported that gold nanoparticles can be used as thermal therapy for tumors and also iron oxide nanoparticles have wide bio medical applications for magnetic resonance imaging, magnetic bio separations and magnetic force-based heating mediators and for drug delivery system [19 and 20]. The above results shows that Fe_2O_3- Au is appropriate for bio imaging and also hold potential for photodynamic therapy.

Acknowledgement
Authors gratefully acknowledge Centre for Research on Molecular and Applied Sciences, Trivandrum for the assistance provided in carrying out the biological study, and SAIF, Cochin for providing facilities for characterization.

References

[1] Kooij ES, Ahmed W, Hellenthal C, Zandvliet HJ, Poelsema B. 2012 Nov 5. From nanorods to nanostars: Tuning the optical properties of gold nanoparticles. *Colloids and Surfaces A: Physicochemical and Engineering Aspects*. **413** 231-8.

.[2] Boisselier E, Astruc D. 2009. Gold nanoparticles in nanomedicine: preparations, imaging, diagnostics, therapies and toxicity. *Chemical socety reviews*. **38(6)** 1759-82.

[3] Sujitha MV, Kannan S. 2013 Feb 1. Green synthesis of gold nanoparticles using Citrus fruits (Citrus limon, Citrus reticulata and Citrus sinensis) aqueous extract and its characterization. *Spectrochimica Acta Part A: Molecular and Biomolecular Spectroscopy*. **102** 15-23.

[4] Teja AS, Koh PY. 2009 Mar 1. Synthesis, properties, and applications of magnetic iron oxide nanoparticles. *Progress in crystal growth and characterization of materials*. **55(1-2)** 22-45.

[5] Agostinis P, Berg K, Cengel KA, Foster TH, Girotti AW, Gollnick SO, Hahn SM, Hamblin MR, Juzeniene A, Kessel D, Korbelik M. 2011 Jul. Photodynamic therapy of cancer: an update. *CA: a cancer journal for clinicians*. **61(4)** 250-81.

[6] Lucky SS, Soo KC, Zhang Y. 2015 Feb 25. Nanoparticles in photodynamic therapy. *Chemical reviews*. **115(4)** 1990-2042.

[7] Davis ME, Chen Z, Shin DM. 2010. Nanoparticle therapeutics: an emerging treatment modality for cancer. *Nanoscience and technology: A collection of reviews from nature journals* 239-50

[8] Reddi E. 1997 Feb 1. Role of delivery vehicles for photosensitizers in the photodynamic therapy of tumours. *Journal of Photochemistry and Photobiology B: Biology*. **37(3)** 189-95.

[9] Larsen GK, Farr W, Hunyadi Murph SE. 2016 Jul 21. Multifunctional Fe2O3–Au nanoparticles with different shapes: enhanced catalysis, photothermal effects, and magnetic recyclability. *The Journal of Physical Chemistry C*. **120(28)** 15162-72.

[10] Kumar LP, George NA. 2020 Jan 20. Enhanced thermal diffusivity of water in the presence of iron oxide nanoparticles: a laser induced thermal lens measurement. *Materials Research Express*. **7(1)** 015064.

[11] Philip D. 2008 Nov 1. Synthesis and spectroscopic characterization of gold nanoparticles.

Spectrochimica Acta Part A: Molecular and Biomolecular Spectroscopy. **71(1)** 80-5.

[12] Rufus A, Sreeju N, Philip D. 2016. Synthesis of biogenic hematite (α-Fe 2 O 3) nanoparticles for antibacterial and nanofluid applications. *RSC advances*. **6(96)** 94206-17.

[13] Lassoued A, Lassoued MS, Dkhil B, Ammar S, Gadri A. 2018 Jul 1. Synthesis, photoluminescence and Magnetic properties of iron oxide (α-Fe2O3) nanoparticles through precipitation or hydrothermal methods. *Physica E: Low-dimensional Systems and Nanostructures*. **101** 212-9.

[14] Kwiatkowski S, Knap B, Przystupski D, Saczko J, Kędzierska E, Knap-Czop K, Kotlińska J, Michel O, Kotowski K, Kulbacka J. 2018 Oct 1. Photodynamic therapy–mechanisms, photosensitizers and combinations. *Biomedicine & pharmacotherapy*. **106** 1098-107.

[15] Luo D, Carter KA, Miranda D, Lovell JF. 2017 Jan. Chemophototherapy: an emerging treatment option for solid tumors. *Advanced Science*. **4(1)** 1600106.

[16] Kharkwal GB, Sharma SK, Huang YY, Dai T, Hamblin MR. 2011 Sep. Photodynamic therapy for infections: clinical applications. *Lasers in surgery and medicine*. **43(7)** 755-67.

[17] F Sperandio F, Huang YY, R Hamblin M. 2013 Aug 1. Antimicrobial photodynamic therapy to kill Gram-negative bacteria. *Recent patents on anti-infective drug discovery*. **8(2)** 108-20.

[18] Talarico LB, Zibetti RG, Faria PC, Scolaro LA, Duarte ME, Noseda MD, Pujol CA, Damonte EB. 2004 Apr 1. Anti-herpes simplex virus activity of sulfated galactans from the red seaweeds Gymnogongrus griffithsiae and Cryptonemia crenulata. *International Journal of Biological Macromolecules*.**34(1-2**) 63-71.

[19] Ghosh P, Han G, De M, Kim CK, Rotello VM. 2008 Aug 17. Gold nanoparticles in delivery applications. *Advanced drug delivery reviews* **60(11)** 1307-15.

[20] Fortin JP, Wilhelm C, Servais J, Ménager C, Bacri JC, Gazeau F. 2007 Mar 7. Size-sorted anionic iron oxide nanomagnets as colloidal mediators for magnetic hyperthermia. *Journal of the American chemical society* **129(9)** 2628-35.

IOP Conf. Series: Materials Science and Engineering 1233 (2022) 012006 doi:10.1088/1757-899X/1233/1/012006

Amplified spontaneous emission from ZnO nanostructures as scatterers in rhodamine 6G

K R Devika[1,*], Merin Joby[1], Frincy Francis[1], C P Jinsi[2], Riju C Issac[2] and Santhi Ani Joseph[1]

[1] Department of Physics and Centre for Research, St. Teresa's College (Autonomous), Ernakulam, Kerala, India

[2] Department of Physics, Cochin University of Science and Technology, Kalamassery, Kochi, Kerala, India

* Corresponding author : devikakr171097@gmail.com

Abstract. Under high power, Nd: YAG 532nm, ns pulsed laser spectral narrowing of an amplified spontaneous emission (ASE) from ZnO nanostructures in rhodamine 6G has been studied. The dependence of emission intensity on excitation power, particle density of ZnO nanostructures, and silica as scatterers show a distinct threshold. The Full width half maximum (FWHM) substantially reduces from 46 nm to 3.16 nm at high excitation power and ZnO-silica scatterer inclusion.

1. Introduction

Amplified spontaneous emission (superfluorescence) is the process where spontaneously emitted photons are amplified by further coherent stimulated emission [1]. In other words, ASE is the amplification of spontaneous fluorescence by stimulated emission. It is readily observed in high gain media like dye solutions when excited using intense light pulses [2]. Amplified spontaneous emission is often regarded as an unwanted effect in laser physics, but it has significant applications like Random lasing. Random lasing is a nonconventional laser whose feedback is formed by random fluctuations of the dielectric constant in space [3]. It can be achieved in a strong gain medium that provides ASE when proper feedback in the form of scattering particles is introduced.

Studies on random lasing in metal oxides like Zinc oxide, Titanium oxide, organic dyes, organic and inorganic nanocomposites etc. have been reported by numerous research groups [4], [5]. [6], [7]. Our interest specifically lies in ZnO owing to its unique characteristics including high bandgap (3.37 eV), larger excitonic binding energy (~60 meV), non-toxic nature etc. The PL spectra of ZnO nanostructures have confirmed distinct emissions at the visible region. *F. Francis et. al* had previously reported the anomalous thermal diffusivity showed by ZnO microtubes and nanoparticles, many folds greater than water and other existing heat transfer fluids. These fascinating properties have triggered great interest in developing ZnO into a self-heat dissipating random laser at the visible region. This work is a first step towards developing a potential random laser out of ZnO finding applications in biological, biochemical, optoelectronic and photovoltaic fields.

This work particularly focused on the behaviour of ZnO nanoparticles as scatterers in rhodamine dye. The study investigated the optimum concentration of ZnO scatterers that give maximum ASE intensity. The effect of silica microspheres as scatterers was also a subject of interest. The threshold magnitudes of certain parameters like pump intensity and dye concentration required to obtain narrowed ASE spikes were achieved by the optimization process.

IOP Conf. Series: Materials Science and Engineering 1233 (2022) 012006 doi:10.1088/1757-899X/1233/1/012006

2. Preparation Technique

ZnO nanoparticles of size 50 nm were synthesized using the low-temperature hydrothermal technique. The hydrothermal method has been opted as it is a simpler and faster synthesis method that enables the control over particle dimensions and, the physical conditions including temperature required for the synthesis process are easily achievable. ZnO nanoparticles were synthesized using Zinc acetate dihydrate [Zn $(CH_3\ COO)_2.2H_2O$] as the Zinc precursor and Sodium Hydroxide [NaOH] as the precipitating agent. These were dissolved separately in deionized water at the desired molarity. The zinc acetate dihydrate was added drop-wise to NaOH solution, then inserted into an electrical oven at 90°C for 2 hours. A white solution with precipitate was formed. The mixture was then centrifuged at 3000 rpm for 30 minutes and the supernatant was removed and the precipitate was cooled to room temperature for 2–4 hours and then washed several times to remove the organic impurities and other unreacted compounds. Later, the precipitate was ground using an agate mortar and pestle for making nanoparticles [8].

3. Characterization

3.1 structural characterization-FESEM

The structural characterization of synthesized samples was done using FESEM using Carl Zeiss Zigma TM instrument with specifications of 5kV and 10X -1.000.000X.

(a) **(b)**

Figure 1. FESEM image of ZnO nanoparticles of size (a)50 nm (b) 100nm

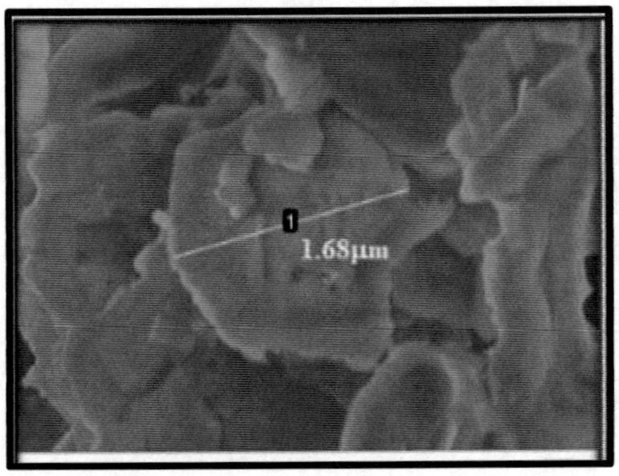

Figure 2. FESEM image of SiO_2 revealing diameter of about 1.68µm

Figure 1 is the SEM image of ZnO nanoparticles of size 50 nm and 100 nm, Figure 2 depicts the SEM image of silica microspheres having a diameter of about 1.68 µm.

3.2 Optical characterization

The optical behaviour was analysed using UV-Visible absorption and PL measurements. To study the absorption spectra, a JASCO V570 UV-VIS-NIR spectrophotometer with a resolution of 1 nm and wavelength range 200-1000 nm was used.

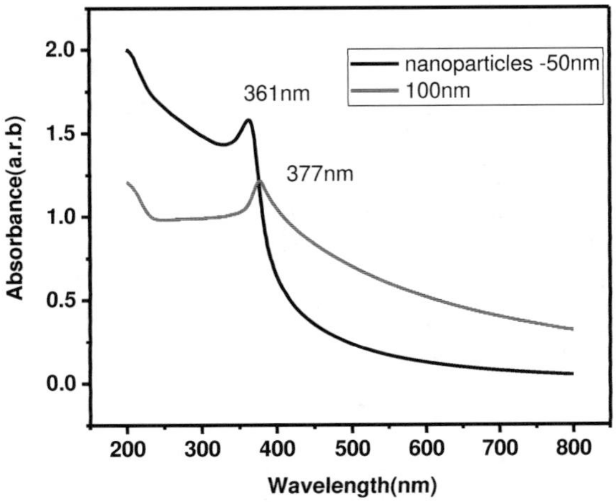

Figure 3. Absorption spectrum of ZnO nanoparticles of sizes 50 nm and 100 nm

Figure 3 shows the absorption spectra of ZnO nanoparticles of size 50 nm and 100 nm with absorption peaks at 361 nm and 377 nm respectively. The PL spectra of ZnO nanoparticles are obtained using Horiba QM 8075 11 C spectrofluorometer with emission range 185-900 nm and resolution of 0.3 nm. The samples were excited at wavelengths 295 nm, 350 nm and 488 nm.

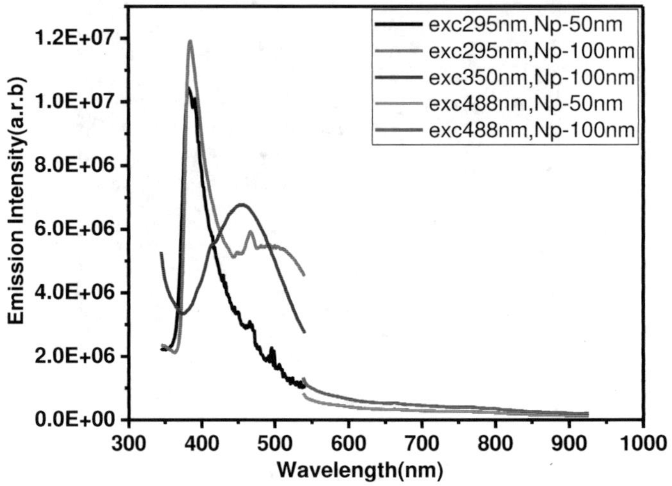

Figure 4. PL of ZnO nanoparticles excited at 295 nm, 350 nm, and 488 nm

Figure 4 shows the PL spectra of ZnO nanoparticles of sizes 50 nm and 100 nm. The spectra show distinct peaks at the visible region apart from the typical ZnO peak at about 391 nm when excited using 295 nm and 350 nm. The broadband emission in the visible region is due to the defect states such as Zn vacancy or O interstitials.

4. Experimental Set up

The difference between normal fluorescence and ASE lies in the excitation source used. An ASE spectrum can be observed when the weak excitation lamp used in normal fluorescence measurements is replaced by a short-pulsed laser. An indication of ASE is the narrowing of the fluorescence spectrum in extension and in time and the substantial elevation in intensity. The study was conducted on Rhodamine: ZnO composites. The absorption wavelength of Rhodamine 6G lies at about 532 nm. Therefore, a frequency-doubled pulsed Nd: YAG laser (532 nm, pulse width -10 ns) was used to observe ASE. The light was focused on the sample using mirror and lens and, the emission was detected and measured using Ocean Optics mini spectrometer and emission was recorded and analysed using Ocean Optics STS-VIS Software.

A stock solution of rhodamine 6G dye having molarity 0.05 mM was prepared. A part of this solution was taken in a four-sided cuvette. The sample was irradiated using a 532 nm pulsed laser and the emission was recorded. Both pump power and dye concentration were varied to determine the optimum ranges for pump intensity and dye concentration for obtaining ASE. Later, ZnO nanostructures were added in equal steps to the optimized rhodamine solution. The ASE intensity is measured in all trials and the optimum particle density of ZnO scatterers for maximum intensity was found out. Silica solution of molarity 1.7 mM was prepared using deionized water. 5 μL of this solution (particle density- 5×10^{12} / cm^3) was micropippetted and added to the optimized rhodamine-ZnO sample to observe changes in emission intensity. The concentration of silica was varied to discover the optimum value.

IOP Conf. Series: Materials Science and Engineering 1233 (2022) 012006 doi:10.1088/1757-899X/1233/1/012006

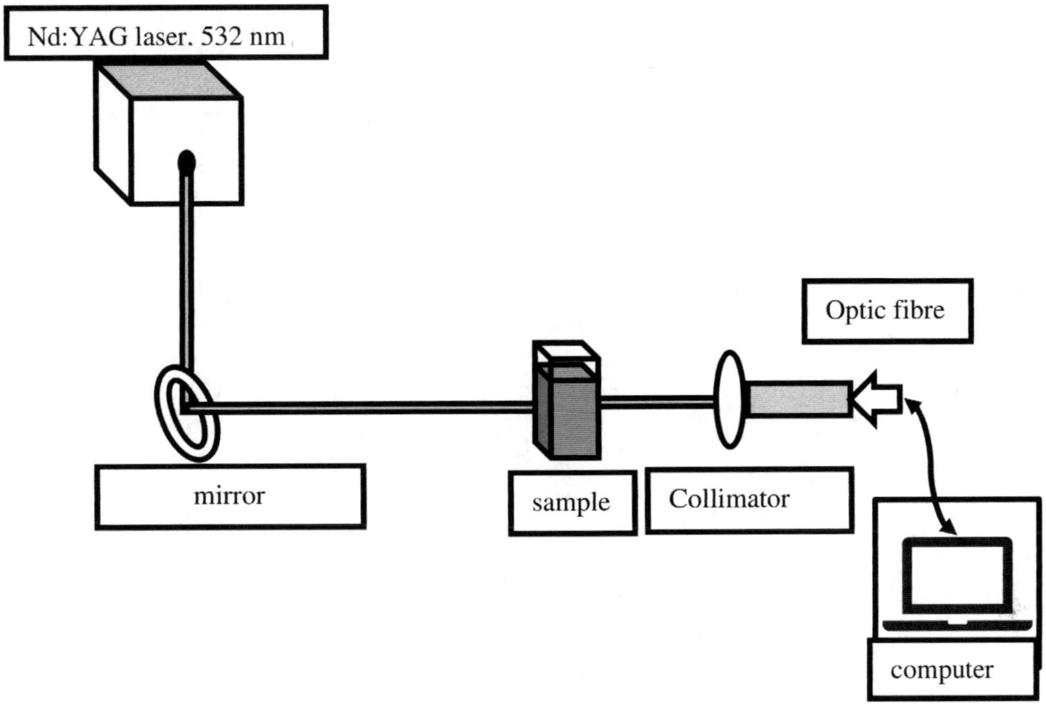

Figure 5. Schematic diagram for experimental set up to measure ASE

Figure 5 shows the schematic diagram for the experimental set-up for measuring ASE. The sample was illuminated with a frequency-doubled pulsed Nd: YAG laser using a mirror. The emitted light was collected using a collimator lens, detected by Ocean Optics mini spectrometer and, the emission was recorded and analysed using Ocean Optics STS-VIS Software.

5. Observations and Results

5.1 ASE measurements

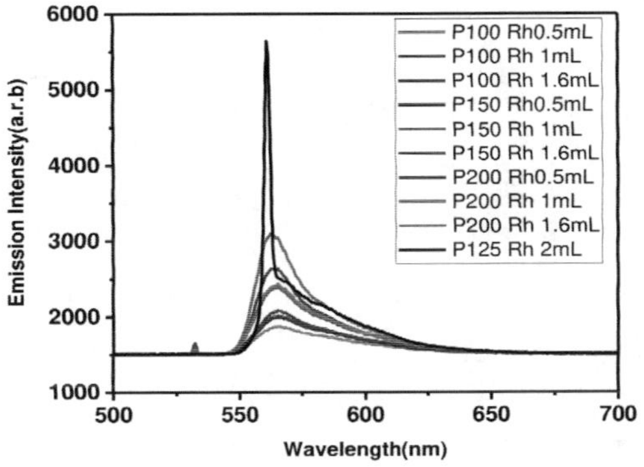

Figure 6. Optimization of pump power using rhodamine dye solution

Figure 6 shows the optimization of dye concentration and pump power. ASE is recorded for different concentrations of dye (0.5 ml, 1 ml, 1.6 ml and 2 ml) at pump powers ranging from 100 mJ to200 mJ. Narrowed and intense ASE was observed at the dye concentration of 2 ml and pump power of 125 mJ.

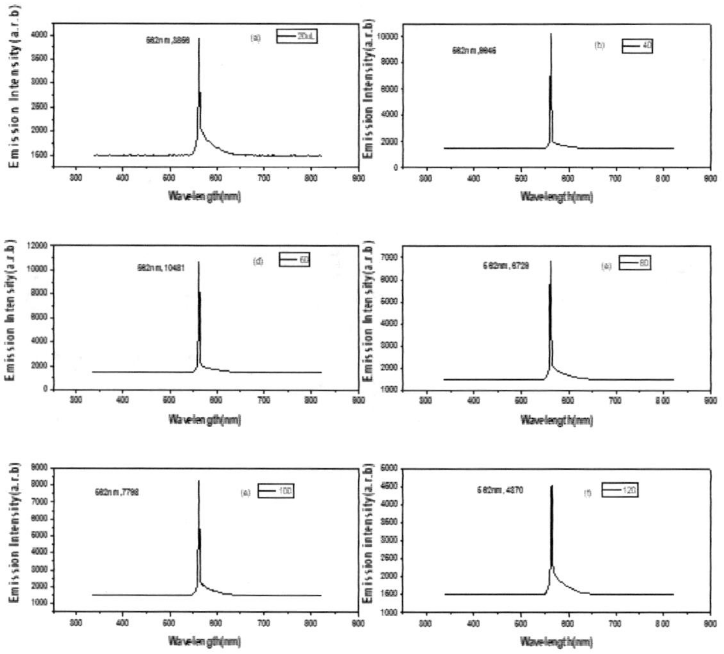

Figure 7. ASE for varying particle density of ZnO nanoparticles of size 50 nm, molarity 1.2 $\times10^{-7}$M, dispersed in 2 ml rhodamine dye Solution, laser power 125mJ(a) 1.4×10^{9} / cm^{3} (b) 2.9×10^{9} / cm^{3} (c) 4.3×10^{9} / cm^{3} (d) 5.8×10^{9} / cm^{3} (e) 7×10^{9} / cm^{3} (f) 8.7×10^{9} / cm^{3}

Figure 7 shows ASE from ZnO nanoparticles of size 50 nm in rhodamine dye with varying particle density. Emission intensity increases with the addition of particles and reaches the maximum particle density of 4.3x 10^{9} / cm^{3} and then decreases. At the threshold particle density, random lasing with a very thin and intense pulse was achieved.

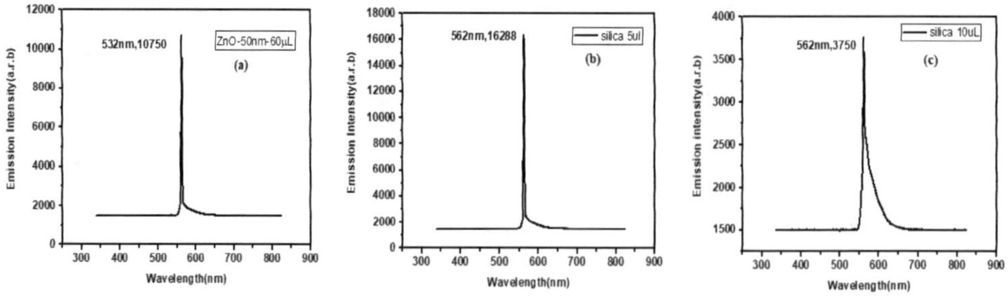

Figure 8. ASE from ZnO nanoparticles of size 50 nm, molarity 1.2 $\times10^{-7}$M, 4.3×10^{9} / cm^{3} particle density (a) without silica (b) with silica particle density-5×10^{12} / cm^{3} (c) 1×10^{13} / cm^{3}

IOP Conf. Series: Materials Science and Engineering 1233 (2022) 012006 doi:10.1088/1757-899X/1233/1/012006

Figure 8 represents ASE intensity from ZnO nanoparticles of size 50 nm, 4.3×10^{9} / cm^{3} particle density with and without silica. The Addition of 5×10^{12} / cm^{3} silica further enhanced particle density and contributed to emission intensity and random lasing. Ideal laser-like emission was observed for ZnO: silica inclusion. But a further increase in particle density tremendously decreased emission and spectral broadening took place.

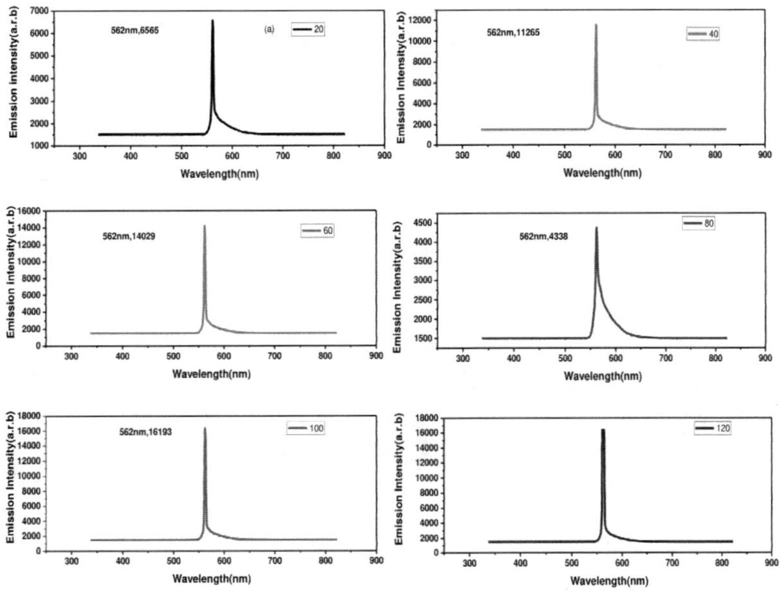

Figure 9. ASE for varying particle density of ZnO nanoparticles of size 100 nm, molarity 1.2 $\times10^{-7}$M, dispersed in 2 ml rhodamine dye Solution, laser power 125 mJ (a) 1.4×10^{9} / cm^{3} (b) 2.9×10^{9} / cm^{3} (c) 4.3×10^{9} / cm^{3} (d) 5.8×10^{9} / cm^{3} (e) 7×10^{9} / cm^{3} (f) 8.7×10^{9} / cm^{3}

Figure 9 shows ASE from rhodamine: ZnO nanocomposite with ZnO nanoparticle size 100 nm at various particle densities. The emission showed enhanced intensity at particle density $7\times10^{9}/cm^{3}$ with a very narrow spectrum.

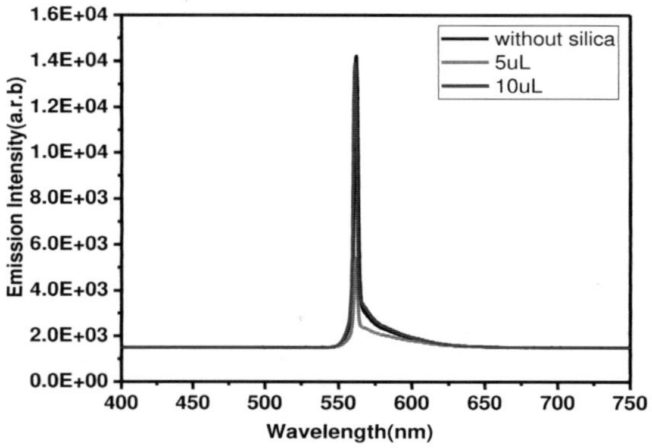

Figure 10. ASE in ZnO nanoparticles, 100 nm size, 1.2 $\times 10^{-7}$M, $7\times10^9/cm^3$ particle density, without and with silica of particle density $5\times10^{12}/cm^3$ and 1×10^{13} / cm^3

Figure 10 shows the ASE intensity of ZnO nanoparticles of size 100 nm with a particle density of 7×10^9 / cm^3 without and with silica. From the plot, it is evident that the addition of silica of particle density $5\times10^{12}/cm^3$ to this sample narrowed the emission spectrum. But further addition caused broadening of the spectrum.

6. Conclusion

The synthesis of ZnO nanoparticles using the low-temperature hydrothermal method yielded mono-dispersed, stable crystalline phased structures. We were able to optimize the conditions for obtaining high intense and sharp ASE pulse from rhodamine dye solution. Initially, a broad spectrum over a wide wavelength range was detected, which later turned to a laser-like emission at about 562 nm when pump intensity and dye concentration were increased. 2 ml of 0.05 mM rhodamine dye solution excited at a pump power of 125 mJ could readily give narrowed pulse as output. The evolution of the emission spectrum confirms the transition from ASE to random lasing.

The challenges that limit applications of random lasers are multidirectional emission, lack of tunability and strong mode competition with chaotic fluctuations. In a strongly scattering random medium, light can be trapped by concurrent scattering leading to the formation of localized cavity modes. Such localization of photons within the cavity formed by scatterers is termed as Anderson localization which aids stable multimode random lasing. The criterion for Anderson localization is that the average distance between random scattering events (localization length) should be shorter than the sample length. Below the lasing threshold, the fluorescence would be damped by unpumped surrounding gain material. When excitation intensity, dye concentration and scatterer concentration increase, light amplification compensates for loss and the lasing threshold could be reached [9]. The addition of ZnO scatterers to rhodamine solution enhanced the scatterer density, thereby reduces localization length shorter than sample length thus affirms Anderson localization.

An optimum particle density for getting maximum random lasing intensity was figured out for ZnO nanoparticles. The optimized concentration for silica scatterers was also determined. A striking enhancement in emission and laser-like narrowing of the spectrum was observed in rhodamine: ZnO nanoparticle: silica composite. We could observe that the emission intensity has almost doubled by the inclusion of $5\times10^{12}/cm^3$ silica microspheres to the optimized concentration of the rhodamine- ZnO nanoparticle ample. The Full width at half maximum (FWHM) substantially reduces from 46 nm to 3.16 nm at optimized conditions and with ZnO-silica scatterer incorporation. The results prove that ZnO nanostructures in rhodamine dye solution can act as nanoresonators that amplify the ASE emission and random lasing intensity. The comparative study between normal fluorescence and amplified spontaneous emission revealed that the peak intensity of ASE was centred around the same wavelength and frequency as that of the peak intensity from normal fluorescence. It is worth mentioning that the inclusion of particles can contribute remarkably towards random lasing and an optimum particle density is existing.

Acknowledgement

Authors gratefully acknowledge DCE-Government of Kerala, DST- WOS-A and DST-FIST for the extended support.

Reference

[1] Dudley, Christopher 2004 Absorption, fluorescence and amplified spontaneous emission of blue-emitting dyes

[2] Penzkofer, Alfons, and Falkenstein W 1978 Theoretical investigation of amplified spontaneous emission with picosecond light pulses in dye solutions *Optical and Quantum Electronics* 10.5: 399-423

[3] Wu X, Fang W, Yamilov A, Chabanov A A, Asatryan A A, Botten L C and Cao H 2006 Random lasing in weakly scattering systems Physical Review A, 74(5) 053812

[4] Cao, Hui, et al. 1999 Random laser action in semiconductor powder *Physical Review Letters* 82.11: 2278.

[5] Xu, C X, et al. 2005 Ultraviolet amplified spontaneous emission from self-organized network of zinc oxide nanofibers *Applied physics letters* 86.1: 011118

[6] Penzkofer, Alfons, and Falkenstein W 1978 Theoretical investigation of amplified spontaneous emission with picosecond light pulses in dye solutions *Optical and Quantum Electronics* 10.5: 399-423

[7] Anglos, Demetrios et al. 2004 Random laser action in organic–inorganic nanocomposites *JOSA B* 21.1: 208-213

[8] Frincy Francis, Anila E I, and Santhi Ani Joseph 2020 Dependence of thermal diffusivity on nanoparticle shape deduced through thermal lens technique taking ZnO nanoparticles and nanorods as inclusions in homogeneous dye solution *Optik* 219: 165210

[9] Liu J, Garcia P D, Ek S, Gregersen N, Suhr T, Schubert M, and Lodahl P 2014 Random nanolasing in the Anderson localized regime nature Nanotechnology 9(4), 285-289.

Facile one pot hydrothermal synthesis of reduced graphene oxide-tin disulfide (rGO-SnS_2) nanohybrid material and their optical and structural investigation

N S Das[1], A Chowdhury[1,2] and A Roy[1*]

[1]Department of Physics, National Institute of Technology Silchar, Silchar, Assam-788010, India.
[2]Department of Condensed Matter Physics and Material Sciences, S. N. Bose National Centre for Basic Sciences, Salt Lake, Kolkata-700106, West Bengal, India.

E-mail: 28.asim@gmail.com

Abstract. Here, we follow a simple, cost effective, and environmentally friendly one step hydrothermal method for the synthesis of nanohybrid material reduced graphene oxide-tin disulfide (rGO-SnS_2). Herein, the spectroscopic and structural investigation of SnS_2 anchored reduced graphene oxide (rGO) such as UV-vis DRS spectroscopy, photoluminescence (PL) spectroscopy, fourier transform infrared spectroscopy (FTIR), and X-ray diffraction (XRD) are studied. A broad peak observes at 365 nm in UV-vis absorption spectra. Optical band gap energy is calculated using Tauc's plot and the estimated value is 1.96 eV. Functional groups such as Sn-S, Sn-O, C = C, -OH, C-H, C-O are identified using FTIR which attributes to the successful preparation of nanohybrid material. Structural analysis is done using XRD. XRD result confirms the hexagonal phase of the nanohybrid material and the peak at $2\theta = 15.19^{\circ}$ confirms the characteristics of SnS_2 and a small peak at $2\theta = 26.03^{\circ}$ indicates the existence of carbon. Interplanar spacing is calculated from Bragg's law. The value is estimated to be 5.82 Å. Average crystallite size of the material is estimated using Scherrer's formula and the value is found to be 28.16 nm. Strain of the material is calculated from Williamson-Hall plot and the value is estimated to be 0.00086.

Keyword: Facile, hydrothermal, Scherrer's formula, nanohybrid

1. Introduction

Among different layered semiconducting materials tin disulfide (SnS_2) has drawn promising attention in modern research field due to its low cost, non-toxic and easily processable [1]. SnS_2 materials have excellent physical, structural and chemical properties which make them promising candidate in various applications [2]. Tin atoms and sulfur anions are bounded by weak van der Waals force and its band gap lies between 2 to 3 eV. Moreover, this material is chemically stable and non-toxic [3]. Owing to high surface area and large number of disordered surface atoms in SnS_2 make them important candidate for high response photo detectors, photocatalysis, gas sensing, electronic and optoelectronic devices [4-7]. Its analogous layered Graphene based materials also attracted growing interest owing to their large surface area, charge carriers, thermal and electrical conductivities [8-10]. These properties make them suitable for anchoring active materials. Recently, intensive research has been done based on nanohybrid material to enhance the superior performance in many fields. Nanohybrid materials are very important in much exploitation such as in photocatalysis, lithium-ion batteries, Sodium ion batteries, supercapacitors, and so forth [3,11-15]. In this article, we try to focus on the optical and structural investigation of hydrothermally synthesized rGO-SnS_2 nanohybrid material. Moreover, this hydrothermal process is absolutely non-toxic, environmentally friendly, low cost and most importantly toxic reducing agents such as sodium borohydrate and hydrazine hydrate are avoided during the synthesis process. During the synthesis, tin source dissociates into ions in the presence of GO solution and it is adsorbed by functional groups of GO.

2. Experimental Section

2.1. Materials and reagents used

Graphite flakes (Alfa Aesar), sodium nitrate ($NaNO_3$) (Merck), sulfuric acid (H_2SO_4) (Merck), hydrochloric acid (HCl) (Merck), potassium permanganate ($KMnO_4$) (Merck), hydrogen peroxide (H_2O_2) (Merck), Stannic chloride ($SnCl_4.5H_2O$) (Loba Chemie Pvt. Ltd.), thiourea (CH_4N_2S) (Merck), DI Water, ethanol.

2.2. Synthesis procedure

Conventional Hummer's method had been used for the preparation of grapheme oxide (GO) [16]. In a typical synthesis of rGO-SnS_2 as schematically displayed in figure 1, pre synthesized GO was added in (1 mg/mL) DI water followed by 1 hour of ultrasonication. After ultrasonic treatment in the bath sonicator, 1.52 g of stannic chloride pentahydrate ($SnCl_4.5H_2O$) had been added slowly into the GO solution and magnetically stirred for 1 hour and added 1.38 g of thiourea (CH_4N_2S) and again kept it for vigorous stirring to get the homogeneous solution. The prepared homogeneous solution was then transferred into a 50 mL Teflon lined stainless steel autoclave. The sealed autoclave was heated at 180 °C in hot air oven for 16 hours. After natural cooling down, the yellow precipitate was cleansed with DI water followed by absolute ethanol. The washed product was dried in hot air oven at 60 °C for 5 hours. Finally, dark yellow coloured rGO-SnS_2 powder was obtained as shown in figure 2.

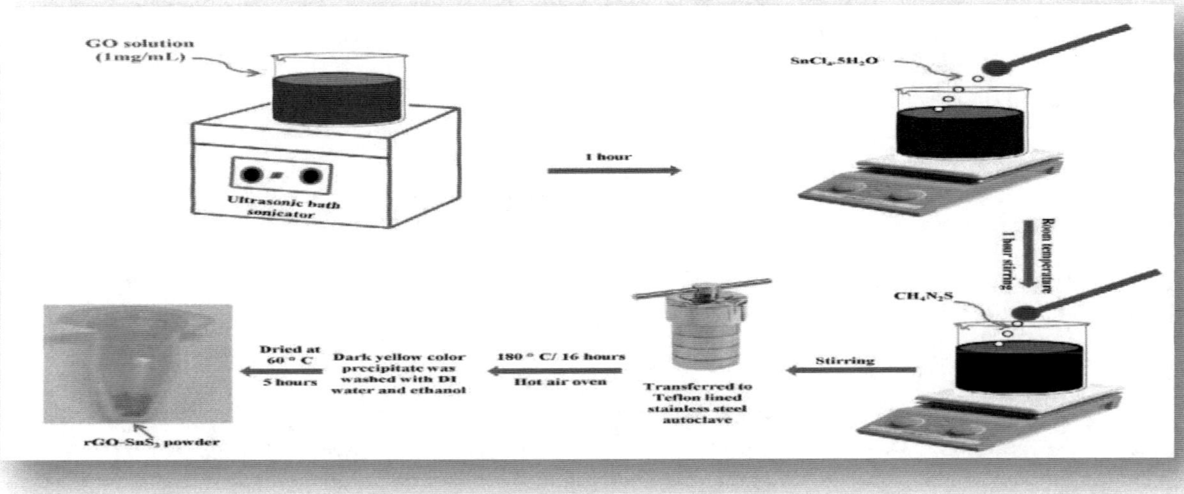

Figure 1. Schematic diagram of synthesis of rGO-SnS_2.

IOP Conf. Series: Materials Science and Engineering 1233 (2022) 012007 doi:10.1088/1757-899X/1233/1/012007

Figure 2. Image of the synthesized rGO-SnS_2 powder.

2.3. Instrumentation

The synthesized product was characterized by Photoluminescence (PL) spectroscopy (Fluoromax-4C Spectrofluorometer, Horiba), FTIR (Nicolet, USA IMPACT 410), X-ray Diffractometer (XRD) (Xpert[3] MRD XL, Panalytical), UV Vis NIR Spectrophotometer (Agilent, Model: Cary 5000).

3. Result and Discussions

3.1. Spectroscopic studies

3.1.1. UV-vis DRS spectroscopy of rGO-SnS_2

Optical characteristics of rGO-SnS_2 is studied using UV-vis DRS spectroscopy as shown in figure 3(a). A wide absorption peak observes at 365 nm. Tauc's plot of the synthesized product is shown in figure 3(b). The values are obtained as 1.96 and 3.91 eV, respectively [17]. The narrow band gap is very much suitable for many applications such as visible light absorption, photocatalytic degradation of dyes etc. [18].

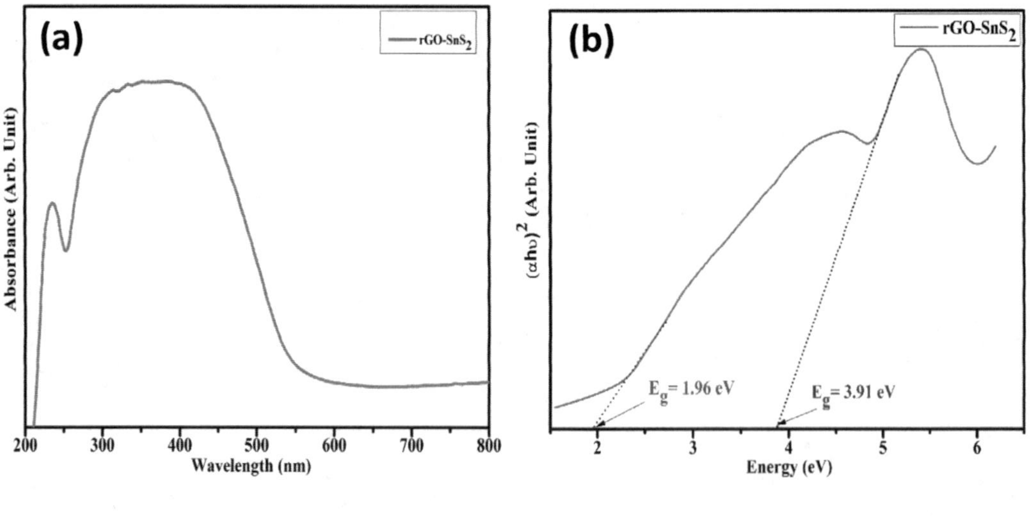

Figure 3. (a) UV-vis DRS spectra and (b) Tauc's plot of rGO-SnS_2.

3.1.2. Photoluminescence (PL) spectroscopy

PL spectrum of rGO-SnS_2 is recorded at (λ_{ex} = 365 nm) as shown in figure 4. It is used to investigate the charge carriers recombination in the material. PL emission spectra exhibits an emission peak at 558 nm that occurs may be due to the absorption of electrons in the higher excited states [18, 19].

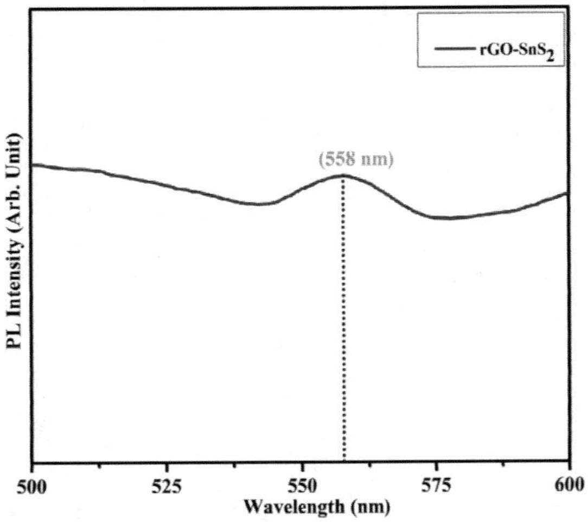

Figure 4. PL spectroscopy of rGO-SnS_2.

3.1.3. Fourier transform infrared (FTIR) spectroscopy

The formation of nanohybrid material is further confirmed by FTIR method. Various well defined peaks have been observed. Peak appears at 3462 cm^{-1} corresponds to the presence of O-H group. 1563 cm^{-1} and the vibrational peak at 1389 cm^{-1} represent the functional groups C = C and C-H, respectively. 1124 cm^{-1} corresponds to C - O group. 659 cm^{-1} and 513 cm^{-1} are attributed to stretching vibration of Sn-S and Sn-O bond, respectively [20, 21] as displayed in figure 5. These peaks suggest the existence of SnS_2 in the nanohybrid. 1563 cm^{-1} represents the vibration of graphene sheets [22]. These results confirm the successful formation of nanohybrid.

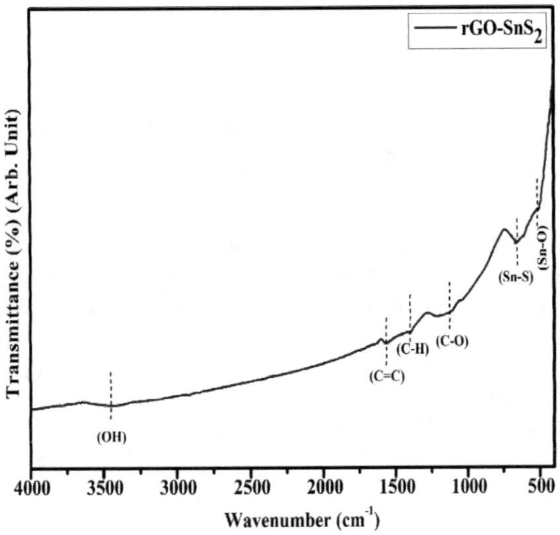

Figure 5. FTIR spectrum of rGO-SnS_2.

3.2. X-ray Diffraction (XRD)

Figure 6 (a) displays the XRD pattern of rGO-SnS_2. The peak at 2θ =15.19 ° is due to the (001) plane of SnS_2 which indicates the characteristics of SnS_2 and a small peak at 2θ = 26.03 ° corresponds to the presence of carbon in the hybrid material [23, 24]. Interplanar spacing of SnS_2 is found out to be 5.82 Å calculated using Bragg's law and average crystallite size of the material is estimated to be 28.16 nm from Scherrer's formula. Williamson-Hall plot provides the strain of the material to be 0.00086 schematically displays in figure 6 (b).

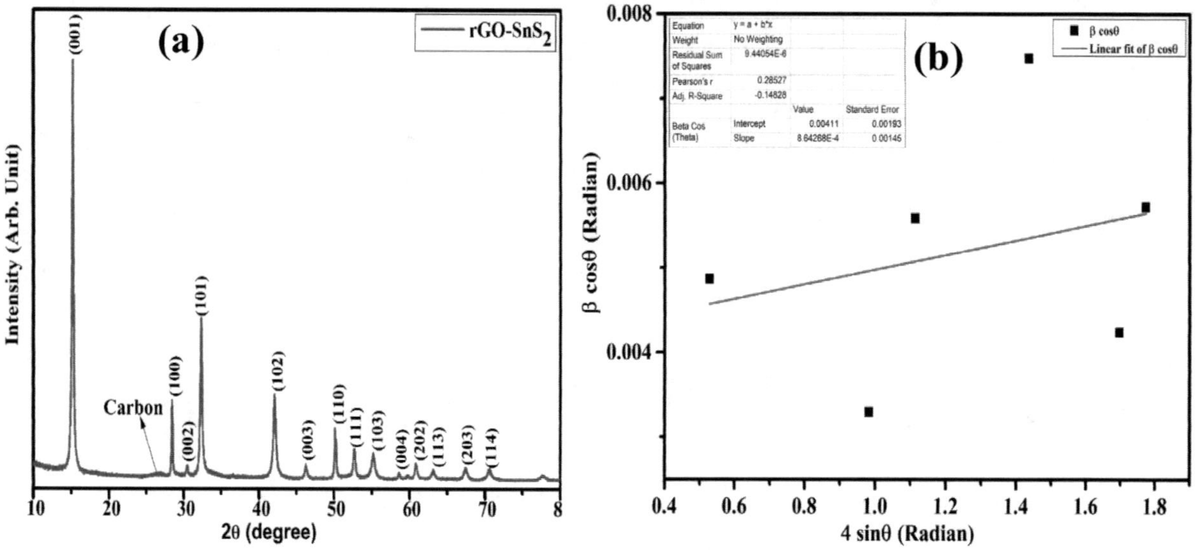

Figure 6. (a) XRD pattern and (b) W-H plot of rGO-SnS_2.

4. Conclusion

In summary, rGO-SnS_2 has been successfully synthesized by facile, low cost and environmentally benign hydrothermal method and avoids harmful reducing reagents like hydrazine hydrate and sodium borohydrate. Metal disulfide (SnS_2) is successfully anchored in rGO during the in-situ synthesis of the nanohybrid material. Optical band gap energies are estimated to be 1.96 and 3.91 eV, respectively from Tauc's plot method. The components and the different functional groups of the nanohybrid material are investigated by FTIR analysis and the presence of functional groups of both SnS_2 and rGO confirms the successful synthesis of the hybrid material. From, XRD analysis interplanar spacing and average crystallite size are estimated using Bragg's law and Scherrer's formula and the values are found to be 5.82 Å and 28.16 nm, respectively. From Williamson-Hall plot, strain of the material is found out to be 0.00086. These cost effective synthesized materials may have found potential applications in new fields of research and technology in future.

5. References

[1] Rout C S, Joshi P D, Kashid R V, Joag D S, More M A, Simbeck A J, Washington M, Nayak S K and Late D J 2014 *Appl. Phys. Lett.* **105** 043109
[2] Duangchuen T, Karaphun A, Wannasen L and Kotutha I 2019 *Applied Surface Science* **487** 634
[3] Rahimi A, Kazeminezhad I and Ghahfarokhi S E M 2018 *Journal of Materials Science: Materials in Electronics* **29** 4449
[4] Sun Y, Cheng H, Gao S, Sun Z, Liu Q, Liu Q, Lei F, Yao T, He J, Wei S and Xie Y 2012 *Angew. Chem. Int. Ed.* **51** 8727
[5] Chao J, Xie Z, Duan X B, Dong Y, Wang Z, Xu J, Liang B, Shan B, Ye J, Chen D and Shen G 2012 *Cryst Eng Comm* **14** 3163
[6] Yang W, Gan L, Li H and Zhai T 2016 *Inorg. Chem. Front.* **3** 433
[7] Li M, Zhu Y, Li T, Lin Y, Cai H, Li S, Ding H, Pan N and Wang X 2018 *Inorg. Chem. Front.***5** 1828
[8] Blanton T N and Majumdar D 2012 *Powder Diffraction* **27** 104
[9] Gupta R K, Alahmed Z A and Yakuphanoglu F 2013 *Materials Letters* **112** 75
[10] Zhu Y, Murali S, Cai W, Li X, Suk J W, Potts J R and Ruoff R S 2010 *Adv. Mater.* **22** 3906
[11] Youn D H, Stauffer S K, Xiao P, Park H, Nam Y, Dolocan A, Henkelman G, Heller A and Mullins C B 2016 *ACS Nano* **10** 10778
[12] Mei L, Xu C, Yang T, Ma J, Chen L, Li Q and Wang T 2013 *J. Mater. Chem. A* **1** 8658
[13] Modarres M H, Lim J H-W and George C 2017 *J. Phys. Chem. C* **121 (24)** 13018
[14] Ma C, Xu J, Alvarado J, Qu B, Somerville J, Lee J Y and Meng Y S 2015 *Chemistry of Materials* **27** 5633
[15] Zhong C, Deng Y, Hu W, Qiao J, Zhang L and Zhang J 2015 Chem. Soc. Rev. 44 7484
[16] Gogoi K K, Das N S and Chowdhury A 2019 *Mater. Res. Express* **6** 085108
[17] Wei R, Zhou T, Hu J and Li J 2014 *Mater. Res. Express* **1** 025018
[18] Dashaiirya L, Sharma M, Basu S and Saha P 2018 *Journal of Alloys and Compounds* **774** 625
[19] Chaki S H, Deshpande M P, Trivedi D P, Tailor J P, Chaudhury M D and Mahato K 2012 *Appl Nanosci* **3** 189
[20] Xu Z, Zhang Z, Gao L, Lin H, Xue L, Zhou Z, Zhou J and Zhuo S 2018 *RSC Adv.* **8** 40252
[21] Wu Z, Xue Y, Zhang Y, Li J and Chen T 2015 *RSC Adv.* **5** 24640
[22] Johra F T and Jung W-G 2015 *Applied Surface Science* **357** 1911
[23] Ji L, Xin H L, Kuykendall T R, Wu S-L, Zheng H, Rao M, Cairns E J, Battaglia V and Zhang Y 2012 *Phys. Chem. Chem. Phys.* **14** 6981
[24] Chauhan H, Singh M K, Kumar P, Hashmi S A and Deka S 2016 *Nanotechnology* **28** 025401

Acknowledgments
Authors wishing to acknowledge CIF NIT Sichar, DST-FIST (No. SR/FST/PSI-212/2016(C)) NIT Silchar, SAIC Tezpur University.

IOP Conf. Series: Materials Science and Engineering 1233 (2022) 012008 doi:10.1088/1757-899X/1233/1/012008

Study of the influence of wire metals on dissimilar welds

M.F. BENLAMNOUAR [1,*], N. BENSAID [1], R. BADJI [1], T. SAADI [1], M. IDDIR [1], M. ZIDANI [2,3], M.A. SOUALILI[1]

[1]Research Center in Industrial Technologies CRTI cheraga 16014, Algiers, Algeria
[2]Laboratoire de Génie Énergétique et Matériaux, BP 145, Biskra 07000 Algérie
[3]Université Batna 2, Faculté de Technologie, BP 53, 05078 Fésdis, Algérie

Abstract. The aim of this study is to investigate the effect wire metal on microstructure and mechanical properties of dissimilar welds between HSLA-X70 high strength steel alloy and 304L austenitic stainless steel produced by automatic tungsten arc welding (TIG). The weld joints were prepared using E304L, E316L, E2209L, and E7010 wire metal. The mechanical characteristics obtained from hardness, tensile and impact testing, were correlated to the optical and SEM microscopy, to establish a relationship between wire metal composition and the microstructures in different weld regions. It is concluded that E2209 wire metal lead to improve in the resilience characteristics and tenacity with a slight reduction in the ultimate tensile strength and hardness.

1. Introduction

High strength low alloys steel (HSLA) and Austenitic stainless steels (ASS) are used by many industries owing to corrosion resistance combined to excellent mechanical properties [1-2]. The most applications of these steels include welding technics that are widely used for joining components in the form of similar or dissimilar welds (DWJ). Currently, DWJ is used to assemble two different materials that have not the same chemical composition. This process can be found in many different fields such as power plants, pipeline and automobile industry [3-5]. Their mechanical properties changes vary sharply from fusion zone to base materials, due mainly to the heterogeneity of microstructures across DWJ. This can consequently provoke the deterioration of DWJ performance [6-7]. Thus, several researches have been pursued to improve DWJ performance. One is to remove welding defects by the evaluation of welding parameters like current and gas flow [8]; the other is improve metallurgical behavior of DWJ by cooling rate and Post Weld Heat Treatment [9-10]. Despite these proposals, DWJ between HSLA and ASS constitutes a major challenge of researchers [7], especially for solving metallurgical problems of fusion zone (FZ) and heat affected zone (HAZ). These problems are concentrated around the fusion line, due to coarsening of microstructures, hardening precipitates and the formation of brittle phases during welding process [9-10]. In view of welding process application, thermal welding cycles provokes the diffusion of some alloying elements like C, Fe, Cr and Ni [10-11]. Hence, carbide precipitations are found mainly in weld interfaces due to higher carbon content of HSLA than that of ASS [11]. It is also observed that Cr and Ni diffuse towards Carbone steel from austenitic steel; however, the diffusion Fe followed an inverse direction [12-13]. This phenomenon is accompanied by the presence of chromium carbides precipitation, intermetallic phases and grain coarsening in HAZ [9-12]. It is important to note that the formation of intermetallic phase precipitation is dominated mainly by sigma phase (σ) in ASS welds [13]. Several studies have been carried out on tungsten arc welding (TIG) to assemblate ASS and low alloy steel

[14-15]. A lot of them concentrate on welding defects and solidification mode of fusion zone [11]. However, there is a lack of investigations that are conducted to study ASS-HSLA steel using different wire metals to improve DWJ characteristic. Therefore, this study aims to improve weld joint characteristic through different chemical compositions of wire metal. The welds were prepared using ASS wire (E308L, E316L), HSLA wire (E7010), and duplex stainless wire metal (E2209) by TIG process. This later is chosen because it is one of the most commonly used welding techniques in industry. Accordingly, an investigation is carried out to show the effect of the main alloying elements on different wire metal (C, Cr and Mo) in metallurgical behavior of 304L-X70 DWJ to satisfy of DWJ reliability for industrial applications.

2. Materials and methods

In this work, the Base materials (304L ASS and HSLA-X70) dimensions were fixed at 100x60x2mm. The diameter of wire metals (E308L, E316L, E2209, and E7010) is adopted and fixed at 1.5 mm to satisfy DWJ penetration. The chemical compositions of base and wire metals, measured by FOUNDRY-MASTER spectrometer, are given in Table 1. Welding process was carried out in perpendicular direction to rolling direction of BM's with square butt joints. Automatic TIG welding tests have been performed to get the same welding parameters for different wire metals based on ASME standard [16]. Standard polishing procedures were used for microstructural observations after welding in such a way that, Glycerine reactif was used with stainless sides for 3min, while the HSLA sides were then etched with Nital 3% solution. The microstructure was characterized by ZEISS optical microscope. Mechanical behavior of DWJ was examined using hardness, Charpy impact and tensile tests. Hardness test was conducted using Vickers scale with a load of 0.2Kg (Hv0.2) perpendicular to welding direction. Charpy impact test (55×5×2 mm) were machined from welded specimens at room temperature [17]. Impact specimens are picked and notched and fractured using an INSTRON Charpy impact machine. Standard tensile specimens were prepared using SATEC INSTRON tensile test machine at a strain rate of 0.1/s, according to ASTM-E8 Standard [18] (Fig.1). Then, fractured surfaces from tensile test are then examined using Gemini SEM 300 microscope.

Table.1. Chemical composition of wire and base materials.

Element %	C	Si	Mn	Cr	Ni	Mo	Cu	Nb	Ti	N	V
HSLA X70	0.12	0.35	1.5	0.03	0.01	0.01	0.01	0.02	0.05	0.01	0.09
304L SS	0.047	0.5	1.85	17.8	9.4	0.2	0.15	0.08	0.01	0.08	-
E308L	0.022	0.6	1.82	18.5	9.5	-	-	-	-	-	-
E316L	0.023	0.6	1.88	18.5	9.6	2.8	-	-	-	-	-
E2209	0.022	0.6	1.84	23.0	9.5	2.8	-	-	-	-	-
E7010	0.12	0.80	0.90	-	-	0.5	-	-	-	-	-

Table.2. Welding process parameters.

Intensity I (A)	Medium voltage MV (V)	Weld speed V (mm/min)	Argon flowrate D(L /min)
72	11.68	2.52	8

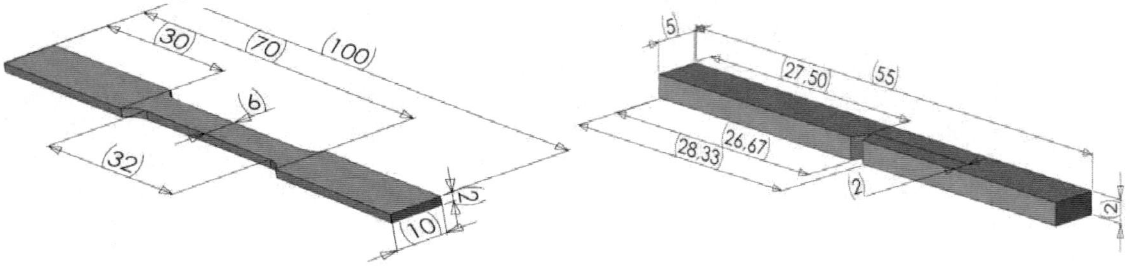

Figure.1 Tensile and Charpy specimens (Left)

3. Results and discussion

3.1. Microstructures evolution

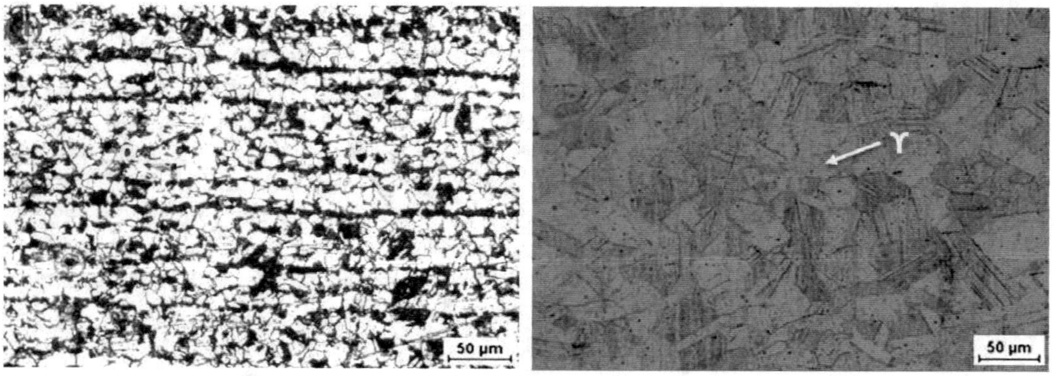

Figure.2 Optical microstructure of the base materials, (a) X70, (b) 304L

Figure 2 shows the microstructure of as received X70 and 304L steels. Rolling microstructure of X70 base material is composed of polygonal ferrite (white color) and pearlite (black color), containing about 75% ferrite with a range of ferrite grain sizes between 14 to 18μm. 304L austenitic stainless steel composed mainly of austenitic phase (about average diameter of 24μm) with a small amount of ferrite phase precipitation in austenite grain boundary. Based on micrographic observations (Fig. 3), it can show that weld joints are free from welding defects like porosity, undercut, and cracks. It distinguishes clearly the fusion zone FZ and weld interfaces on both parts of base materials. The microstructures of weld joint can be divided into three sections: fusion zone (FZ) and heat affected zones (HAZ-304L, HAZ-X70).

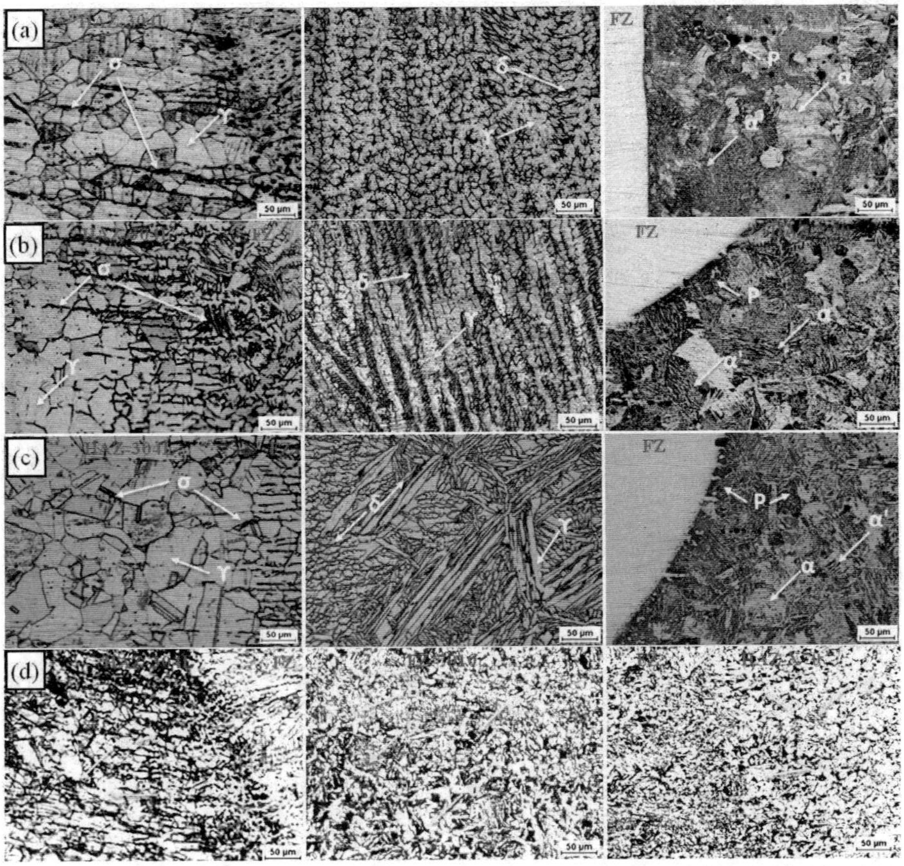

Figure.3 Weld microstructures, (a) 304L weld, (b) 316L weld (c) 2209 weld, (d) 7010 weld

During TIG welding process, DWJ undergoes metallurgical transformations due to the welding thermal cycle (WTC). Due to the high gradient of heat input through the WTC, HAZ is produced in both sides of the base materials adjacent to the fusion zone (HAZ-X70, HAZ-304L). HAZ-304L undergoes an increasing in austenite grain size when becoming close to FZ (Fig. 3). This coarsening is enormous, when using 308L (Fig. 3a).

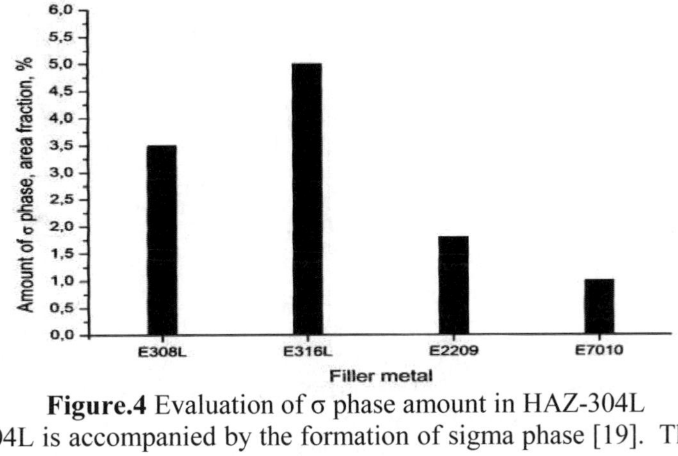

Figure.4 Evaluation of σ phase amount in HAZ-304L

In addition, HAZ-304L is accompanied by the formation of sigma phase [19]. This intetmetalic phase is resulted from the decomposition of delta-ferrite to sigma phase over ferritic transformation. It is indicated by intermittent and thick lines, and identified with lamellar morphology organized in tetragonal crystal structure [18-12]. Furthermore, sigma phase variations can be assigned either to welding thermal cycles and alloying element variations of the wire metal. Thus, wire metal compositions contribute to a variation in the proportion of sigma precipitate phase at high temperatures. This variation is mainly related to the C, Cr, Ni and Mo content (Table 1). Thus, the molybdenum of 316L provoke an increase of sigma phase proportion in HAZ-304L (Fig. 3b) compared to 308L (Fig. 3a, Fig. 4), while the higher chromium rate of E2209 results a decreasing in sigma phase proportion, which is closer to 1.8% for HAZ-304L (Fig. 3c, Fig. 4). The Carbon content of E7010 wire metal also has a main effect on metallurgical behavior of DWJ. This induces a slight decrease of sigma phase rate and a increasing of carbide precipitation compared the other wire metals. Moreover, this carbide precipitation appears in the form of dark spots over HAZ-304L/FZ boundary (Fig. 3d). The HAZ-X70 microstructures are mainly formed by acicular ferrite and pearlite, containing large ferrite content compared to X70 base material (Fig. 3a-d). These microstructures are distinguished by the formation of small amount of martensite slats; this amount is even greater in the samples welded by E7010 (Fig. 3d). In stainless fusion zone (E308L, E304L, E2209), ferritic transformation is occurred due to high welding thermal cycles [11] to get austenitic matrix, the ferritic solidification mode leads to form austenite phase surrounded by residual ferrite (δ) under different morphologies, resulting from the changes of wire metal composition. Therefore, E308L fusion zone (FZ-308L) consists of austenitic phase and ferrite (δ). It is characterized by vermicular morphology and δ ferrite fraction of 7.5% (Fig.3a). The FZ-316L is represented by lathy morphology with δ amount of 11.5% (Fig.3b). Furthermore, FZ-2209 exhibits needles morphology in by where the δ ferrite is around 22.5% (Fig.3c). However, the FZ-7010 is partially quenched, it is obtained by a rapid cooling of the wire metal, contains a higher rate of martensite (α'), bainite (B), acicular ferrite, pearlite and a small portion of carbide precipitation (Fig.3d).

3.2. Mechanical behavior

Vickers hardness of welded samples is shown in Figure 5a. A steady hardness is obtained for the base materials. BM-X70 hardness is relatively higher than BM-304L as expected by the microstructure and the chemical composition (Table 1).

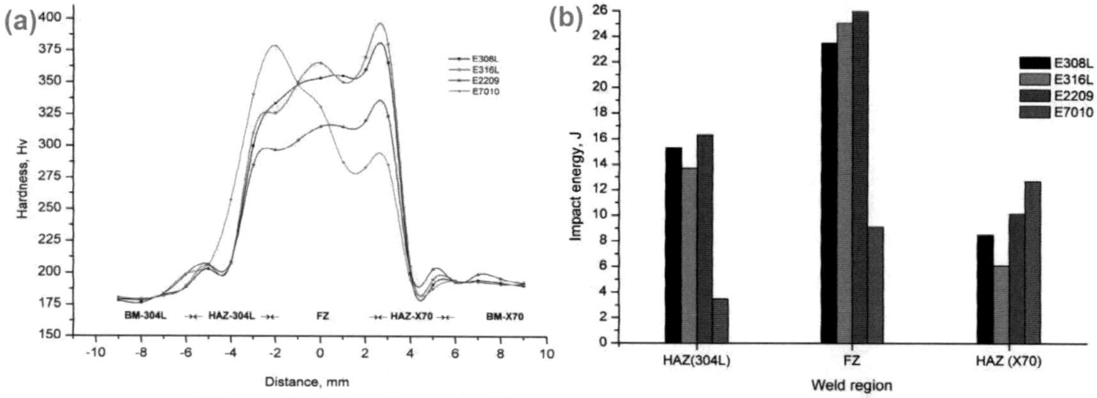

Figure.5 (a) Hardness profile, (b) Impact energy of weld region

Hardness increases is observed in the heat affected zone (HAZ's) of both base materials, (Fig.5a). These higher values are related to presence of intermetalic phase (σ) in HAZ-304L and the formation of hard phase (α') in X70 base material. Higher increases in hardness peaks were observed between HAZ-X70 and FZ, due to high amount of carbide precipitation produced in fusion line boundary (Fig. 3). This phenomenon is also observed in HAZ-304L. Hardness values are seriously perturbed in FZ-7010, this phenomenon can be explained by their complex microstructure which constituted mainly of ferrite and martensite lattices. It is observed that FZ-308L and FZ-316L produce approximate values

of hardness due to similar microstructures which is formed by these two austenitic wire metals. The FZ-2209 exhibits a balanced level of hardness compared to the other wire metals.
Figure 5.b. presents impact energies of the HAZ's and FZ of weld specimens. It indicates that E2209 gives the better combination between impact resistances (K) of three weld regions compared to the other samples. However, the lower energy was found in HAZ-304L welded by E7010 (HAZ-304L/7010). This can be explained by the forte heterogeneity between stainless side (HAZ304L, FZ-308L, FZ-3016L, FZ-2209) and HSLA side (HAZ-X70, E2209). Furthermore, it is observed that homogeneous bimetallic joints have excellent impact properties (low Carbone fusion zone with X70 base materials and stainless fusion zones with 304L base material) compared to the heterogeneous joints.

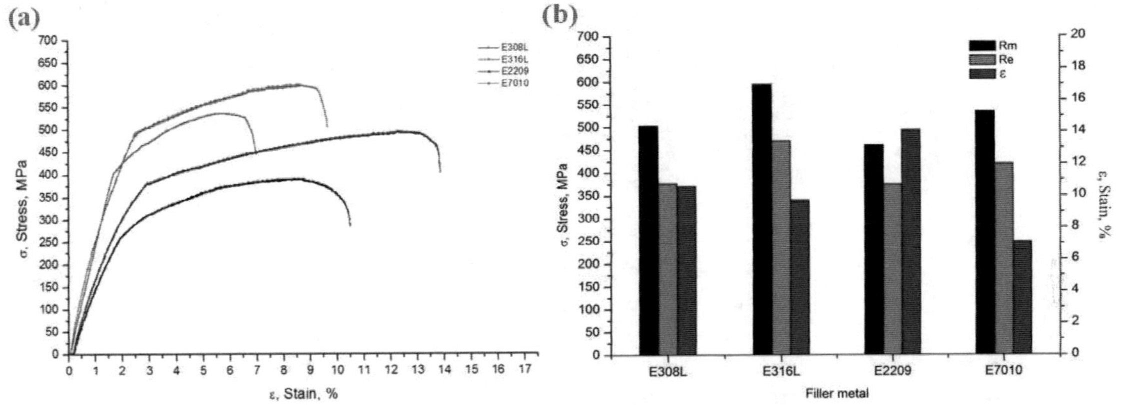

Figure.6 Tensile test of weld joint, (a) tensile curve, (b) tensile characteristics
Ultimate strength and maximum strain of welded specimens are shown in figure 6a. It observed that tensile properties changes significantly in terms of the chemical composition of wire metal. This is attributed by both the rate and the nature of phase in weld regions (FZ, HAZ). Thus, stain and elongation measurement are mainly influenced, by FZ morphology, the amount of precipitate phases (δ, γ), metastable phases (α', B), austenitic and ferritic grain growth in HAZ's. Among the different tensile testes (fig.6b), the E2209 wire metal clearly gives a better combination of stain and ductility compared to the other wire metals.
Fractography evolution
The Examination of the fracture surfaces by tensile impact test is illustrated in Fig.7.
It notes that, the tensile fracture is located in the HAZ in such a way that E308L, E316L and 2209 provokes the fracture in HAZ-X70. However, when using E7010, fracture is found in HAZ-304L. This phenomenon is strongly related to both the higher hardness of fractured region and the homogeneity of the chemical compositions between base and wire metal. In fractured HAZ-X70 with E308L (HAZ-X70/308L), the HAZ presents two kinds of surfaces. There exists smooth surfaces indicate a quick propagation of cracking with brittle fracture aspect, and rough surfaces indicate a slow propagation. However, the fracture of HAZ-X70/316L exhibited transgranular morphology, which contains clefts and river forms, with the presence of several crack arrests. The HAZ-X70/2209 is distinguished by ductile fracture with granular aspect, which corresponds to a significant ratio of micro voids and high rate of plastic deformation. Brittle cleavage processes are observed in HAZ-304L/7010, which is persisted with a small plastic deformations (ε %) (Fig.6a), where it occurs by decohesion between crystalline planes, and the surfaces produce smooth striations characterized by cracks developed on parallel lines.

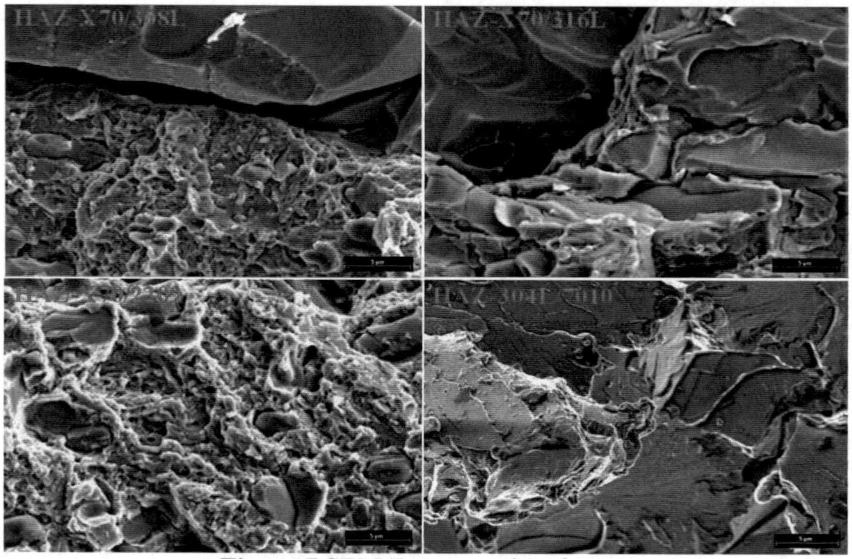

Figure.7 SEM Fractographs of HAZ

4. Conclusion

This study investigates the effects of wire metal types on 304L SS/ HSLA DWJ by the use of different wire metal. Based on the results obtained, it was possible to conclude that:

Mechanical properties of dissimilar weld joints are strongly depend of wire metal composition. In weld joint, there was an increase in the hardness of both FZ and HAZ-X70 compared to HAZ-304L. This corresponds to a significant decrease in mechanical properties of FZ and HAZ-X70 Compared with the HAZ-304L. In the case of E316L and E308L, There is a proportional relationship between the sigma percentage and molybdenum content. This percentage can be decreased by Cr addition in E2209 case. DWJ of 304L SS/ HSLA-X70 has a good weldability with E2209 duplex wire metal compared to austenitic wire metals and low Carbone wire metal by medium energies of TIG process. It gives a butter combination between tensile proprieties and impact resistance. Thus, E2209 welds exhibit the best grain refinement in heat affected zone which contains a smalle amount of σ precipitation phases in 304L heat affected zone that provokes a ductile fracture by tensile test.

To understand more realistically the application DWJ in industry, electrochemical tests will be very important to evaluate weld resistance against corrosion degradations.

Références

[1] M. Ouadah, O. Touhami, R. Ibtiouen, M.F. Benlamnouar, M. Zergoug. Electrical Power and Energy Systems, Vol. 91, 2017, p. 34–41.

[2] N. Naveen Kumar, G. D. Janaki Ram, S. S. Bhattacharya1, H. C. Dey, S. K. Albert. Transactions of the Indian Institute of Metals, Vol. 68, No. 2, 2015, p 289-S297.

[3] S. Ningshem, M. Sakairi. Journal of Solid State Electrochemistry, Vol. 19, No. 2015, p. 3533-3542.

[4] M.F. Benlamnouar, M. Hadji, R. Badji, N. Bensaid, T. Saadi, Y. Laib dit Laksir, S. Senouci: Solid State Phenomena, Vol. 297, 2019, p. 51-61.

[5] N. Bensaid, M. Hadji, R. Badji, M.F. Benlamnouar, T. Saadi, Y. Laib dit Laksir, S. Senouci: Solid State Phenomena, Vol. 297, 2019, p. 195-203.

[6] K. Fan, G.Z. Wang, F.Z. Xuan, S.T. Tu: Engineering Fracture Mechanics, Vol. 136, 2015, p. 279-291.

[7] M. R. A. Shawon , F. Gulshan, A. S. W. Kurny: Journal of The Institution of Engineers (India): Series B. Vol. 96, No. 01, 2015, p. 29-36.

[8] N. Ghosh, P. K. Pal, G. Nandi: Engineering Science and Technology, Vol. 20, No. 04, 2017, p. 1334-1341.

[9] Le Thi Nhung, Pham Mai Khanh, Le Minh Hai, Nguyen Duong Nam: Acta Metallurgica Slovaca, Vol. 23, No. 4, 2017, p. 363-370.
[10] Fanny Mas, et al: Materials Science & Engineering, Vol. 667, 2016, p. 156-170.
[11] H. Tasalloti, P. Kah, J. Martikainen: International Journal of Advanced Manufacturing Technology, Vol. 71, 2017, p. 363-370.
[12] N. Arivazhagan, Surendra Singh, Satya Prakash, G.M. Reddy: Materials and Design, Vol. 32, 2011, p. 3036-3050.
[13] N. Venkateswara Rao, G. Madhusudhan Reddy, S. Nagarjuna: Materials and Design, Vol. 32, 2011, p. 2496-2506.
[14] M F. Benlamnouar, T. Saadi, N. Bensaid, M. Gousmine, Y. Touggui, M. Temmar : Lecture Notes in Mechanical Engineering-Springer, 2021, pp. 463-472.
[15] T. Saadi, M F. Benlamnouar, N. Bensaid, A. Boutaghane, M A. Soualili, H. Hachemi: Defect and Diffusion Forum .2021Volume 406. pp. 319-333.
[16] ASME: An International Code of ASME Boiler and Pressure Vessel Committee on welding and Brazing, American Society of Mechanical Engineers, USA, 2010.
[17] ISO 148. Teste d'impact sur échantillons ISO en V. NF EN ISO 148-2, France, 2009.
[18] ASTM: Standard test methods for tension testing of metallic materials. American National Standard, E8/E8M, 2010.
[19] Chih-Chun Hsieh, Tao-Chih Chang, Dong-Yih Lin, Ming-Che Chen, Weite Wu: Metals and Materials International, Vol. 13, No. 5, 2007, pp. 411-416.

IOP Conf. Series: Materials Science and Engineering 1233 (2022) 012009 doi:10.1088/1757-899X/1233/1/012009

Iron oxide nanoparticles mediated hyperthermia on cancer cell lines

T Samuel Reinhard, Umme Hani Khanum and Elcey C Daniel*

Biotechnology Research Centre, Department of Life Sciences, Kristu Jayanti College (Autonomous), K. Narayanapura, Kothnur Post, Bangalore- 560077 India

***elcey@kristujayanti.com**

Abstract. Nanotechnology is the engineering of functional systems at the molecular scale. The technology can be used to treat cancer by the application of hyperthermia to magnetic nanoparticles and is found to be effective in cancer repression. Iron oxide nanoparticles were synthesized by co-precipitation method. Synthesized nanoparticles were ranging from 10 to 20nm size with Iron concentration of 6 µg. Cancer cell lines, A549 and HeLa, were subjected to hyperthermia treatment and the effect of iron oxide nanoparticles was analysed by Microscopic and Spectroscopic analysis, along with MTT assay in order to check the toxicity of the nanoparticles. Viability of the cells with respect to the concentration of the nanoparticles was determined. Under the influence of the electromagnetic field, iron oxide nanoparticles reduced the viability of the cancer cell lines *ie*., 44% after 5 minutes, 31% and 29% after 10 minutes of exposure.
Key words: Nanoparticles, Cell lines, Hyperthermia, Iron oxide

1. Introduction

Recent and rapid advancement in nanotechnology has improved the ability to design and manufacture the biomaterials for biomedical application. Nanoparticles (NPs) have been used in various fields of medicine and physics. Its utility even during the Roman period for dye fabrication, colouring glass and treatment of arthritis, was well documented [1]. The use of nanoparticle (NPs) materials offers a lot of advantages because of their unique size and physical properties [2]. Though the nanoparticles are the small object of any element, they act as a whole unit with respect to its transport and properties. Some of the common NPs are carbon-based NPs (Fullerenes and carbon nanotubes (CNTs)), metal NPs (Cu, Ag and Au), ceramic NPs, semiconductor NPs, polymeric NPs, lipid-based NPs etc. The NPs have

IOP Conf. Series: Materials Science and Engineering 1233 (2022) 012009 doi:10.1088/1757-899X/1233/1/012009

been synthesized by different methods and they are classified based upon their morphology, size, and chemical properties. Iron oxide nanoparticles (FeO NPs) consist of maghemite or magnetite particles with a diameter ranging from 1-100 nm. They have a high binding capacity and excellent dispersibility property in the solution. This nanoparticle shows varied interest in potential biomedical application due to superparamagnetic properties. Repression of the tumour was found to be effective with magnetic nanoparticle associated hyperthermia [3]. Magnetic properties and biocompatibility of these nanoparticles enhance its use as a potential mediator for cancer treatment [4] . The synthesis of uniform-sized nanoparticle with controllable sizes is very important because the properties of these particles depend on their dimensions. The biomedical application also requires the nanoparticles which have high magnetization values, with sizes smaller than 100 nm, and a narrow particle size distribution. The optical property of the iron oxide nanoparticles is one of the important characteristics for the evaluation of its optical and photo catalytic activity. The outer electrons of atoms or molecules absorb radiant energy through UV/visible absorption and undergo transitions to high energy levels [5]

The magnetic nature of the nanoparticle is due to 4 unpaired electrons in the 3^d shell. When an external magnetic field is applied to a ferromagnetic material, magnetization (M) increases with the strength of the magnetic field (H) until it approaches saturation. The magnetic Iron oxide Nanoparticles exhibit magnetic properties and thus obeys coulombs law of electrostatic force of attraction. The magneto-crystalline anisotropy energy (EA) in the nanoparticles depends on the super paramagnetic properties of ferrite nanoparticles and its size. This energy barrier influences the magnetization direction [6]. According to the Stoner-Wohlfarth theory, EA of a single domain particle is approximated by EA = KVsin2q, where K is the magneto-crystalline anisotropy constant, V is the volume of the nanoparticles, and q is the angle between the direction of field-induced magnetization and the easy axis of the nanoparticles. Magnetic nanoparticles (MNPs) have been extensively studied and investigated for various applications. These particles possess high surface energies because of their large surface volume interactions. Naked NPs have high chemical activity and get easily oxidized in air, generally resulting in loss of its magnetism and dispersibility. Methods for preparing iron oxide magnetic nanoparticles include micro-emulsion, thermal decomposition, and co-precipitation method. As one convenient and economical method, chemical co-precipitation has the potential to meet the increasing demand for the direct preparation of well-dispersed nanoparticles [7].

Size-dependent properties of the nanoparticle are used for both imaging and development of efficient drug delivery system in treatments of cancer. Advanced treatment procedures adopt different techniques by conjugating nanoparticles with antibodies, drugs, proteins, enzymes, and nucleotides which can be specifically directed into organ, tissue, or tumour using an external magnetic field or can be heated in alternating magnetic fields for use in hyperthermia [8]. The magnetic Iron oxide nanoparticles, as such have been used for imaging in MRI (magnetic resonance imaging) scans and also in detection tool for molecular diagnosis by conjugating MNPs with different antibodies and molecular markers [9]. The particles are manipulated by their properties using magnetic field gradient in response to the alternating current magnetic field. Thus it produces the energy transfer effect characterizing the magnetic hysteresis which results in hyperthermia leading to cellular death in cancerous cells [10].

Magnetic hyperthermia generates heat [11] by the Iron Oxide Nanoparticles in response to the external alternating magnetic field resulting in the loss of thermal energy to the environment facilitated by the hysteresis loop produced by the magnetic field [12]. Based on the mechanism of ferromagnetic nanoparticle when administered to the tumour site under alternative magnetic field (AMF), the magnetic-mediated hyperthermia can lead to the killing of tumour tissue or cells without affecting the normal cells [13] [14]. Cancer cells divide rapidly or uncontrolled cell division facilitating solid tumours with the abnormal cell as they are less specialized than normal cells, thus continue to divide without stopping. These cells have an influence on the immune system to stay alive and grow [15]. The growth and repair of body tissues are categorized by the process of programmed cell division. However, the incidence of metastasis can be initiated by mutation and epimutation of the genetic material of the normal cell [16].

Hyperthermia is an age-old technique in the treatment of various diseases. The use of hyperthermia began as a therapy using varied temperature in the tumour mass between 41 and 45°C by external physical means [17] [18]. The applied elevated temperature in the complex tumour-mass has to be kept homogeneous to prevent further multiplication or complete destruction. By incorporating the MNPs in tumour tissues, the temperature exerted by the external alternative magnetic field can be kept uniform which may lead to the declining of cells.

2. Methods

Iron nanoparticles for the study were synthesized in Biotechnology Research Centre at Kristu Jayanti College, Bangalore. Analytical grade chemicals were used for the synthesis of MNPs, determination of Iron concentration and sample preparation. Ferrous sulphate heptahydrate ($FeSO_4.7H_2O$), Ferric chloride hexahydrate ($FeCl_3.6H_2O$) Potassium thiocyanate (KSCN), Hydrochloric acid (HCl), Hydrogen peroxide (H_2O_2), Tetramethyl ammonium hydroxide ($C_4H_{13}NO$), N,N-Dimethylformamide (C_3H_7NO), formaldehyde (CH_2O, 41%) and Acetone (C_3H_6O) were procured respectively from HI Media Laboratories, Thomas Baker Chemicals Pvt. Ltd. and SD Fine-Chem Limited, Mumbai, India. Magnets with the strength of 30 and 50 Gauss were used for creating a magnetic field and separation of particles during washing process.

2.1. Synthesis of magnetic nanoparticles

The magnetic nanoparticle was synthesized by mixing Ferrous sulfate and Ferric chloride in nitrogenized water (1:1 ratio) and added with an equal volume of Tetra-methyl ammonium hydroxide leading to formation of black coloured fluid. One volume of acetone was mixed with 10µl of the above solution and mixed well to obtain a flat shaped pellet. The pellet was dried at 60° C for 30 minutes and re-suspended with acetone. This pellet was dissolved with nitrogenized formaldehyde followed by distilled water and made up the volume with nitrogenized formaldehyde.

The synthesized particles were measured for Iron concentration by potassium thiocyanate method [19]. The samples were diluted with 300µl 6N HCl containing 1% H_2O_2 to obtain the oxidized ferric state of Iron. Potassium thiocyanate solution (5%) was added to the sample to form the coloured complex of Fe III with the thiocyanate and absorbance was measured at 480 nm.

$Fe^{3+}{}_{(aq)} + SCN^-{}_{(aq)} \rightarrow [FeSCN]^{2+}{}_{(aq)}$

The hyperthermia instrument was designed and constructed for the Analysis.

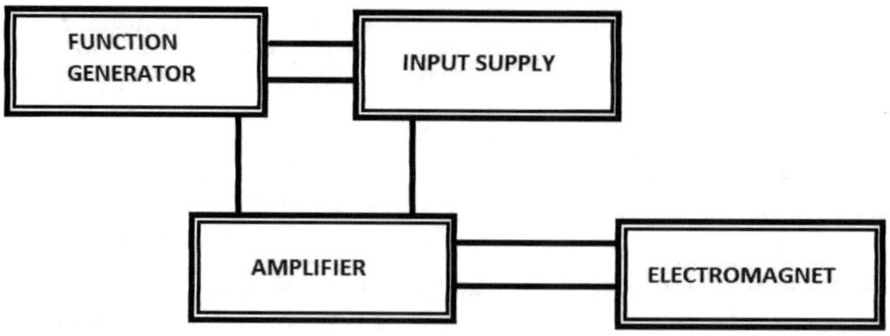

Figure 1: Block diagram of the hyperthermia instrument

2.2. Hyperthermia analysis

Suspensions of cell lines, A541 Lung cancer and HeLa cervical cancer cell line, 200µl each was grown in a 6 well plate to get the required density per well. The plates were incubated for 24 hr at 37°C in a

5% CO_2 atmosphere. The Iron oxide nanoparticles (150 μg) were added to each plate. The plates were placed on an electromagnet and subjected to hyperthermia. The electromagnet was supplied with 400 kHz and 12V for durations of 5 and 10 min. The spent media was removed and MTT reagent was added to a final concentration of 0.5mg/mL of total volume. The plates were incubated in dark condition for 3 hours. MTT reagent was substituted with 100 μl of DMSO, a solubilization solution. Absorbance was read on ELISA reader at 570nm and 630nm used as the reference wavelength. The synthesized particles were characterized by UV Vis Spectroscopy, FTIR, SEM-EDS. Toxicity of NPs and viability of cell lines after hyperthermia were determined by MTT assay [2].

3. RESULTS

The nanoparticles were synthesized using the co-precipitation method. Effect of hyperthermia on Lung and cervical cancer cell lines was analyzed using synthesized iron oxide nanoparticles.

The concentration of Iron in the synthesized particles was determined using the colorimeter as it formed the red complex with thiocyanate. The concentration of iron in sample was measured as 6μg per 100 ml. The result confirmed that the particle formed was Iron nanoparticles. The surface morphologies of the prepared samples were studied using a Scanning Electron Microscope.

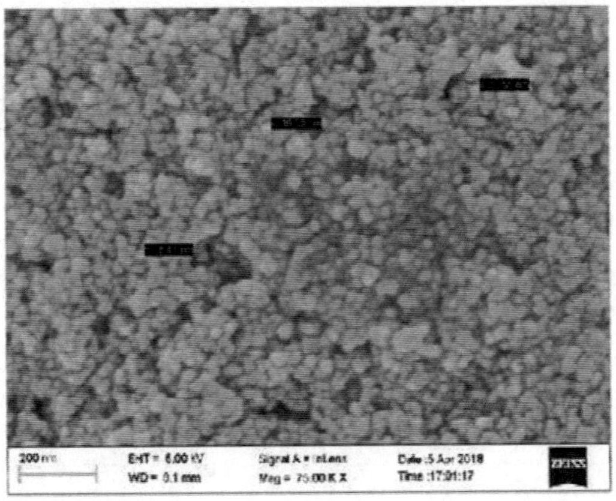

Figure 2: Microscopic image of the iron oxide nanoparticles

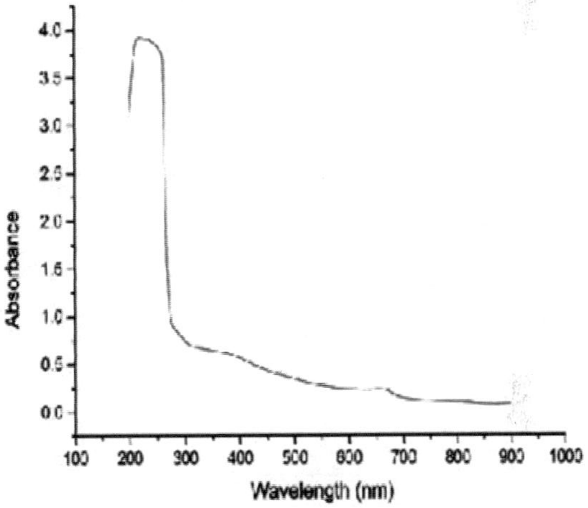

Figure 3. UV Vis spectroscopy iron oxide nanoparticles

The sizes of the synthesized Iron oxide nanoparticles were ranging from 10 nm to 20 nm in size and spherical in nature (Figure 2). The sample was found to be in clumped without any agglomeration. The spherical nature of the NPs states that the NPs are a less toxic compound in comparison to cuboidal shape.

The absorbance of the synthesized NPs on UV Vis Spectroscopy showed spectral ranges between 250 and 300 nm (Figure 3). Among the peaks, the maximum absorbance was found at 230nm, is the characteristic peaks for SPIONs. This value agrees with the earlier report where the nanoparticles exhibited peak absorbance of 250nm [20]. The FTIR spectrum of particles manifested the prominent absorption band located between 3433 and 2121 is considered to the region of iron-oxide nanoparticle in comparison with the range of 3384 and 2073 [21].

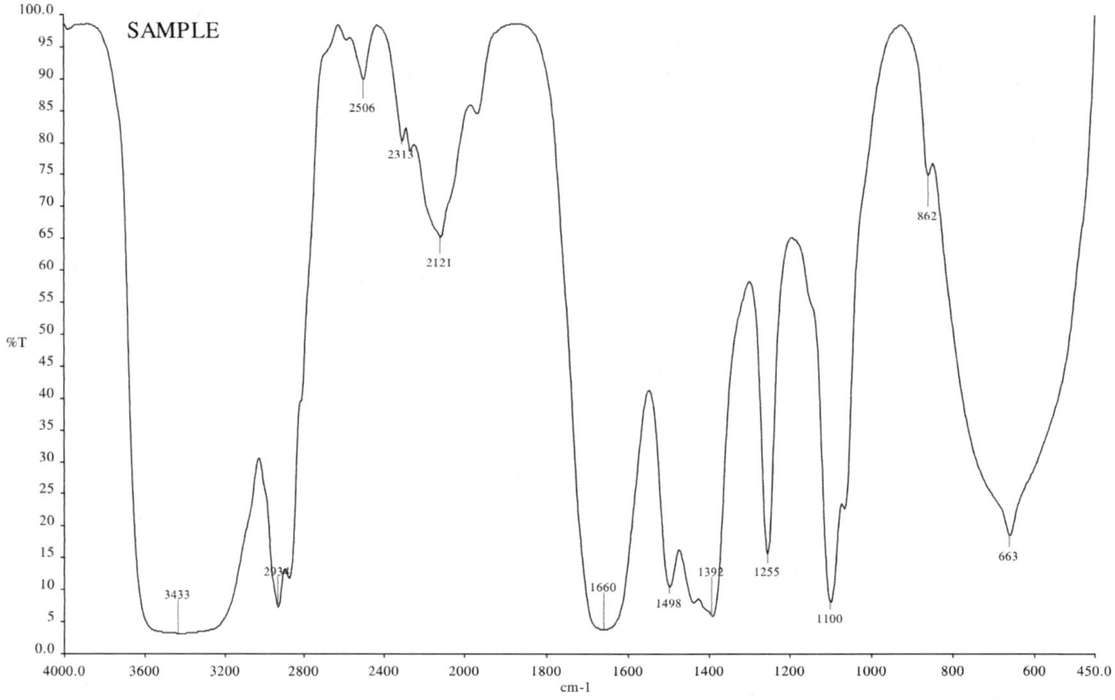

Figure 4: FTIR analysis of naked nanoparticles mixed with NN-dimethylforamide

Table 1: Characteristics IR absorption frequency of functional group present in iron oxide nanoparticle mixed with NN-dimethylforamide.

Absorptions	**Functional Group**	**Type of vibration**	**Intensity**
3433	OH	stretch, H-bonded	strong, broad
3384	OH	stretch, H-bonded	strong, broad
2121	C≡C	stretch	variable, not present in symmetrical alkynes
2073	C-N	stretch	strong
2934	C-H	stretch	strong
2506	OH	stretch	strong, very broad
1680	C=O	stretch	strong
1498	C=C	stretch	medium-weak, multiple bands
1392	C-F	stretch	strong
1255	C-O	stretch	strong
1100	C-O	stretch	strong
663	C-Cl	stretch	strong

The bond types are O-H and -C≡C- (Figure 4) where, former bond is freely bonded, strong and sharp compare to the later, a variable type which is a characteristics of an organic compounds present in the sample. The rest of the peaks depict the functional group of NN-dimethyl formamide present along with the NPs (Table 1).

The MTT was performed to identify the toxicity level of the synthesized nanoparticle against different cell lines, A549 lung cancer and HeLa cervical cancer cell line. The cell lines, A549 and HeLa, were treated with different concentrations of iron-oxide nanoparticle i.e. 50µg, 150µg, 250µg. The following results were observed as in table 2. The toxicity of the nanoparticles against cell lines increased with its increased concentrations. Although the toxicity effect on the cell was less compared to the positive control camptothecin (CPT), where it killed the 51 % of cells cultured. The HeLa cell lines showed the least viability of 61% at 250 µg concentrations of the NPs and high viability was observed at a concentration of 50 µg with 90 % of viable cells.

Table 2. MTT assay results nanoparticles vs A549 and HeLa cell lines

	Control	**Concentration of nanoparticles (µg) vs Cell lines**					
	CPT 25 µg	**50**		**150**		**250**	
		A549	HeLa	A549	HeLa	A549	HeLa
OD value of viable cells	0.40±0.003	0.68 ± 0.002	0.62±0.006	0.60± 0.008	0.52±0.002	0.51±0.005	0.46±0.008
Standard Error	0.0025	0.0015	0.0045	0.006	0.0155	0.0040	0.0060
Viability %	51.66	100	90.82	87.76	73.34	74.48	62.76

The least viability of A549 (table 2) was observed at 250 µg with 74.47% of cell viability and highest viability at 50 µg with 100 % viability of cells. The IC_{50} value was found to be 475µg (Figure 5&6).

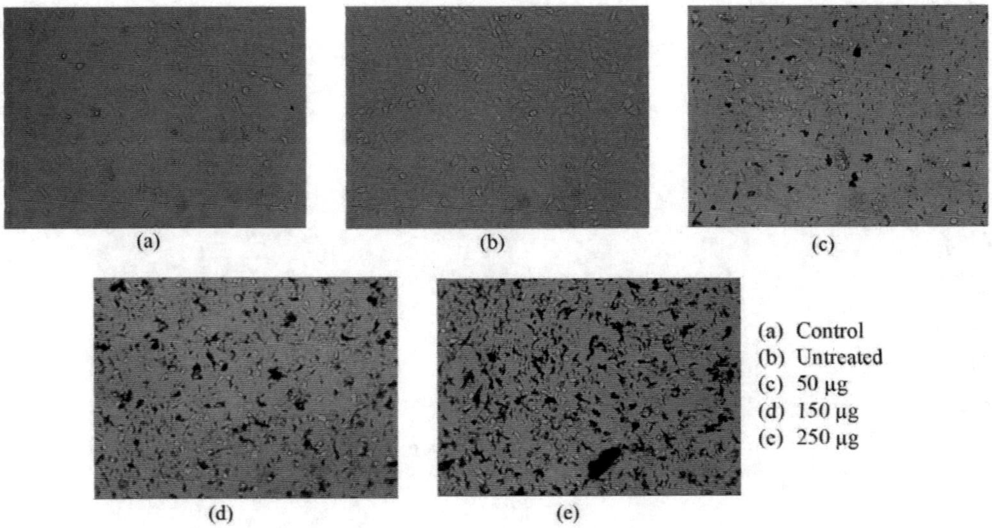

Figure 5: MTT assay results of Cell lines, A549, treated with different concentration of Iron nanoparticles (a) control (b) untreated (c) 50µg (d) 150 µg (e) 250 µg.

The cell lines, A549 and HeLa, were subjected to hyperthermia for the duration of 5 and 10 min respectively. Based on IC_{50} value, 150µg of nanoparticle was used for hyperthermia. Viability of each

IOP Conf. Series: Materials Science and Engineering 1233 (2022) 012009 doi:10.1088/1757-899X/1233/1/012009

cell lines was measured after subjecting to hyperthermia. The result showed that the viability of A549 cell line was 44% and 31% after hyperthermia duration of 5 and 10 minutes respectively (Figure 7).

Whereas the viability of the HeLa cells was found as 44% and 29 % after 5 and 10 min of hyperthermia respectively. The viability of both the cell lines was found to be similar in 5 min of hyperthermia whereas at 10 minutes duration the difference in viability was recorded with 2% less in HeLa cell lines (Figures 8-11).

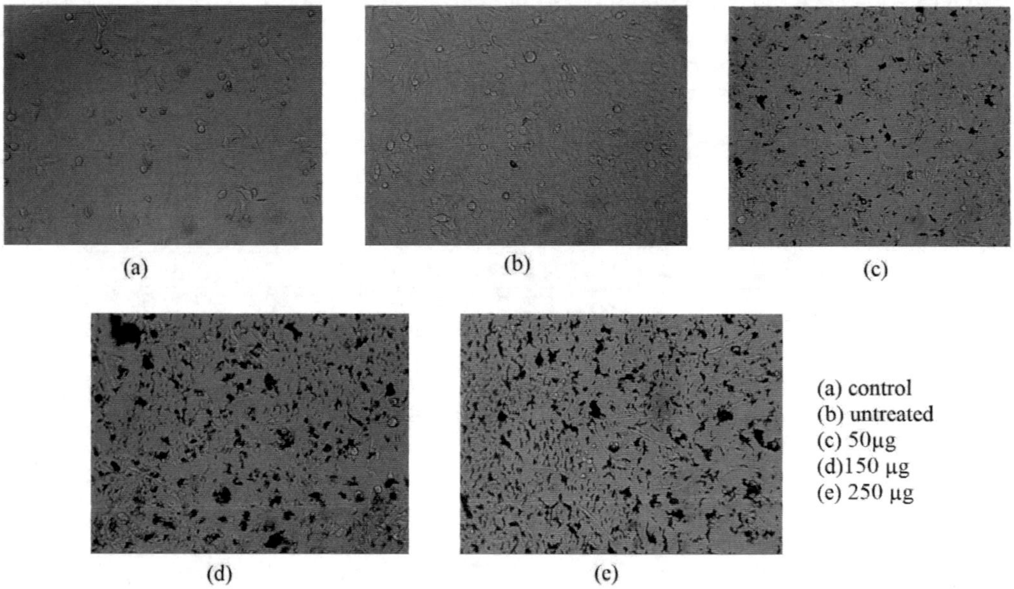

Figure 6: MTT assay results of Cell lines, HeLa treated with different concentration of Iron nanoparticles (a) control (b) untreated (c) 50μg (d) 150 μg (e) 250 μg

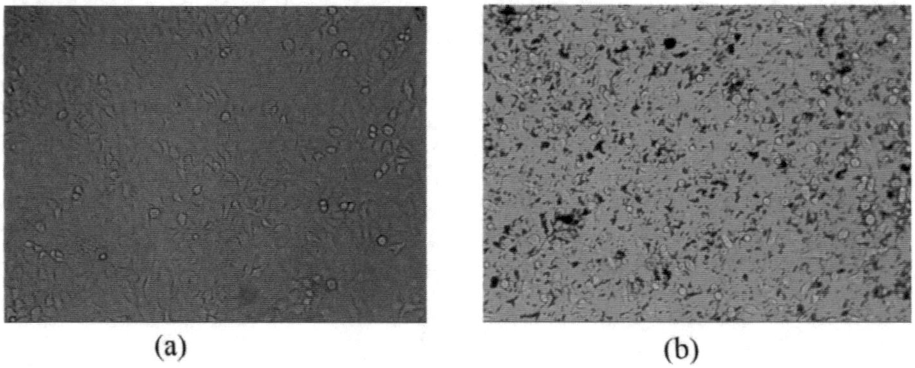

Figure 7: Images of A549 cell line before and after hyperthermia exposure of 5min (a) before and (b) after the exposure

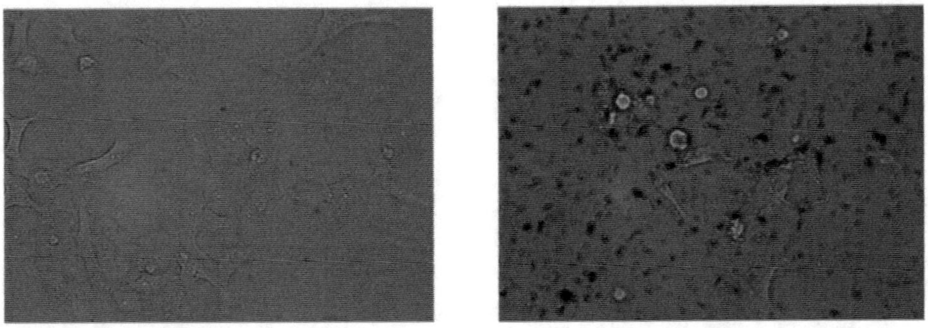

(a) (b)

Figure 8: Image of HeLa cell line before and after hyperthermia exposure of 5 min
(a) before (b) after the exposure

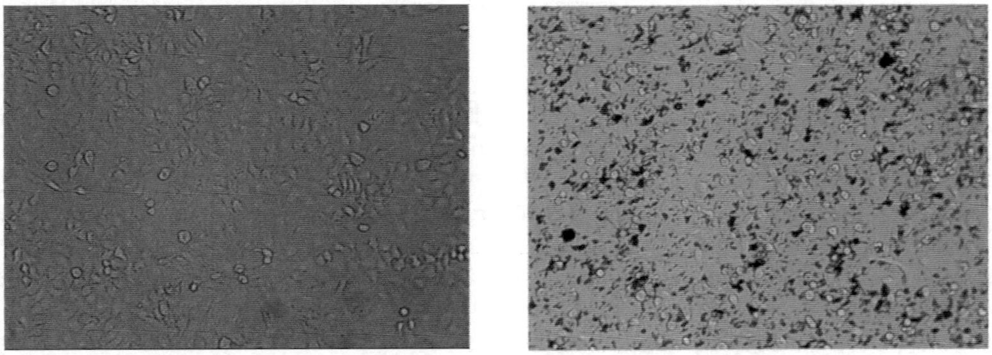

(a) (b)

Figure 9: Image of A549 cell line before and after hyperthermia exposure of 10min
(a) before and (b) after the exposure

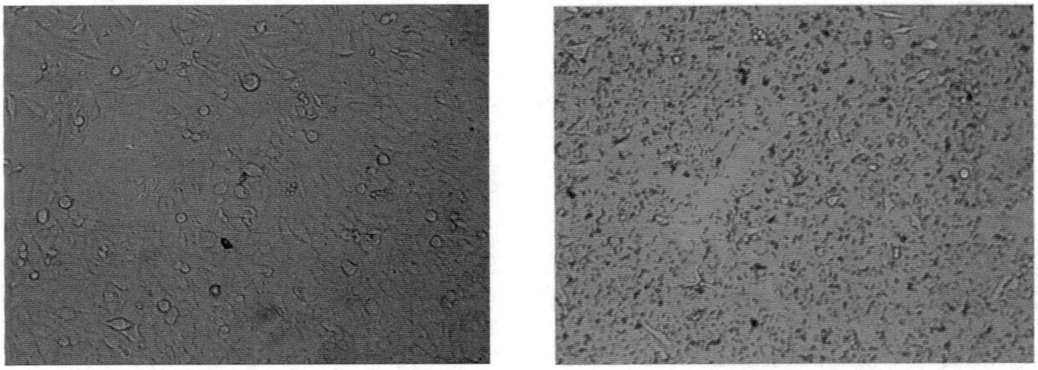

(a) (b)

Figure 10: Image of HeLa cell line before and after hyperthermia exposure of 10min
(a) before and (b) after the exposure

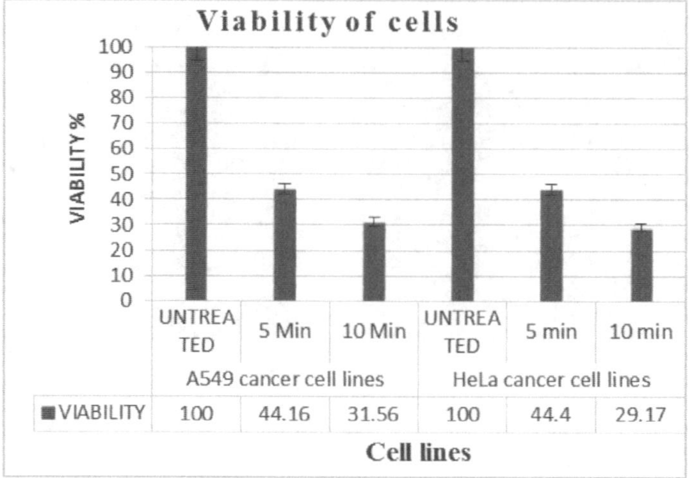

Figure 11: Graph showing the viability % of A549 and HeLa cell lines

4. DISCUSSION

Nanotechnology has proved to be effective in delivering the specific drug at the targeted site using nanoparticles which results in the dissipation of energy in the form of heat under the electromagnetic field. It leads to the conversion of electrical energy to heat energy by the influence of electromagnetic field causing disruption of the cells by the leakage of cell proteins and denaturation of specific enzymes, resulting in the death of cancer cells. Iron oxide Nanoparticles have varied interest in biomedical application due to their super paramagnetic properties. Hyperthermia is considered to be one mode of treatment for cancer associated with the traditional therapeutic approaches such as induction of electromagnetic field which facilitates the agitation of magnetic nanoparticles. Due to agitation it promotes the dissipation of energy in the form of heat thus resulting in the killing of cancer cells. Iron oxide nanoparticles mediated hyperthermia has the potential to achieve localized tumour healing without any side effects. The study involves targeting of magnetic nanoparticles to tumour tissues followed by application of alternating magnetic field that induces the generation of heat thus associated with polar and non-polar repulsion of nanoparticles. The temperature in tumour tissues increases up to 40-45°C which causes necrosis of cancer cells but does not damage healthy surrounding tissues.

Evaluation of efficiency of hyperthermia using iron oxide nanoparticles on cell lines, human cervical cancer (HeLa) and human epithelial cancer (A549), revealed a regression in cell line survival. This is can be explained as they were efficient in generating heat and resulting in the progressive killing of cancer cell lines. Hyperthermia generate in the electromagnetic field with the influence of Nano particles is called as magneto-hyperthermia which had two notable affects. Firstly the treatment showed intense apoptosis and fibrosis in the tumour cells which was detected using MTT assay and visualized by using inverted microscopy. Secondly the treatment showed the utilization of iron oxide nanoparticles in biomedical remedy depends on the shape and surface charges of the nanoparticles influencing their biological distribution. The naked nanoparticles showed its effectiveness due to the adaptation of Para magnetism to the cell surface of the cancer cell line.

The surface morphology of Nanoparticles was studied using Scanning Electron Microscopy which revealed that the Nanoparticles were ranging from 10-20nm in size and appeared to be spherical in shape which impart less toxicity. The UV-VIS spectroscopy showed the absorbance range between 250-300nm confirm to be Iron oxide nanoparticles. The FTIR spectrum showed the prominent absorption band located between 3433 and 2121 is considered to be the region for Iron oxide

Nanoparticles and also shows the presence of O-H and -C≡C- as the functional group of the molecule. Toxicity level of the synthesized nanoparticles were carried out by the MTT assay using HeLa and A549 cancer cell lines which showed 61% of viability at 250μg and 91% of viability at the concentration of 50 and 100μg of nanoparticles respectively. Concentration of 150μg was used to perform hyperthermia, the viability ofA549 cell lines were found to be 44% after 5 minutes and 31% after 10 minutes of exposure to hyperthermia. Whereas the HeLa cell lines showed the viability of 44% and 29% respectively after 5 and 10minutes exposure. The study depicts the successful delivery of nanoparticles into the cells under optimized conditions and effective killing of cancer cells when subjected to hyperthermia under the influence of the electromagnetic field. It also highlights the effectiveness of hyperthermia against cancer cell lines over conventional method of treatment of cancer.

5. REFERENCES

[1] Rao CNR. Nano world – an introduction nanoworld- an introduction to nano science and technology. 2010; *Nava Karnataka Publication Pvt Ltd. ISBN- 13:9788184672473, P 3-5.*

[2] Faraji M, Yamini Y, Rezaee M Magnetic Nanoparticles: Synthesis, Stabilization, Functionalization, Characterization, and Applications. *J. Iran. Chem. Soc.* 2010; *7:* 1–37.

[3] Mukherjee A, Castanares M, Hedayati M, Wabler M, Trock B, Kulkarni P, Rodriguez R, Getzenberg RH, DeWeese TL, Ivkov R, Lupold SE. Monitoring nanoparticle-mediated cellular hyperthermia with a high-sensitivity biosensor. *Nanomedicine 2014; 9(18): 2729–2743. doi:10.2217/nnm.13.207.*

[4] Teja SA, Koh PY. Synthesis, properties, and applications of magnetic iron oxide nanoparticles. Progress in crystal growth and characterization of materials. 2009; 55(1-2): 22-45

[5] Horikoshi & Serpone, 2013 Horikoshi S and Serpone N. Microwaves in Nanoparticle Synthesis, First Edition, *Wiley-VCH Verlag GmbH & Co. KGaA 2013 ISBN: 978-3-527-33197-0*

[6] Martinez *et al.,* 2015 Martinez A, Garcia M, Perry, DL (2015). Study of the properties of iron oxide nanostructures. *Research in Nanotechnology Developments ISBN: 978-1-60741-028-7*

[7] Sun et al., 2010. Sun J, Zhou S, Hou P, Yang Y, Weng J, Li X, Li M. Synthesis and characterization of biocompatible Fe3O4 nanoparticles. *J. Biomed. Mater. Res. Part. A* 2006, *10*: 333–341.

[8] DeNardo *et al.,* 2005 DeNardo SJ, DeNardo GL, Miers LA, Natarajan A, Foreman AR, Gruettner C, Adamson GN, Ivkov R. Development of Tumour Targeting Bioprobes (111In-

Chimeric L6 Monoclonal Antibody Nanoparticles) for Alternating Magnetic Field Cancer Therapy. *Clinical Cancer Research 2005; 11 (19): 7087s-7092s.*

[9] Gupta and Gupta,2005. Gupta AK, Gupta M. Synthesis and surface engineering of iron oxide Nanoparticles for biomedical remediation's. *Biomaterials 2005; 26(18): 3995-4021.*
[10] Liu et al., 2007 Liu Z, Cai W, He L, Nakayama N, Chen K, Sun X, Chen X, Dai H. In vivo biodistribution and highly efficient tumour targeting of carbon nanotubes in mice. *Nat Nanotechnol. 2007; 2:47–52.*

[11] Tietze *et al.,* 2012 Tietze R, Lyer S, Dürr S, Alexiou C Nanoparticles for cancer therapy using magnetic forces. *Nanomedicines. 2012; 7(3): 447-57.*

[12] Giustini, 2010 Giustini AJ, Petryk AA, Cassim SM, Tate JA, Baker I, Hoopes JP. Magnetic Nanoparticle Hyperthermia In Cancer Treatment. *Nano Life. 2010; 1(01n02)* doi: 10.1142/S1793984410000067.

[13] Amin, 2014; Hergt *et al.,* 2006 Amin MZ. Application of Electromagnetic Waves in Cancer Treatment by Hyperthermia. *2014J Electromagnetic Analysis and Applications 6: 209-212.*

[14] Hergt *et al.,* 2006 Hergt R, Dutz S, M¨uller R, Zeisberger M Magnetic particle hyperthermia: nanoparticle magnetism and materials development for cancer therapy, Journal of physics: condensed matter *2006; 18(38): S2919- S2934.*

[15] Shewach and Kuchta, 2009 Shewach DS and Kuchta RD. Introduction to cancer chemotherapeutics. *Chem Rev. 2009: 109 (7):2859-2861.*

[16] Rydzanicz *et al.*, 2006 Rydzanicz M, Golusinski P, Kuchta MD, Golusinski W, Szyfter K, Cyclin D1 gene (CCND1) polymorphism and the risk of squamous cell carcinoma of the larynx. *Archiv für Klinische und Experimentelle Ohren- Nasen- und Kehlkopfheilkunde 2006; 263(1):43-8 https://doi.org/10.1007/s00405-005-0957-7*

[17] Martinez *et al.,* 2013, Martinez-Boubeta C, Simeonidis K, Makridis A, Angelakeris M, Iglesias O, Guardia P, Cabot A, Yedra L, Estradé S, Peiró F, Saghi Z, Midgley PA, Conde-Leborán I, Serantes D, Baldomir D. Learning from nature to improve the heat generation of iron oxide Nanoparticles for magnetic hyperthermia applications *Sci Rep. 2013;3:1652*

[18] Heirgiest *et al.,* 1999 Hiegeister R, Andra W, Buske N, Hergt R, Hilger I, Richter U, Kaiser U. Application of magnetite ferrofluids for hyperthermia. *J Magnetism and Magnetic materials 1999; 201(1-3): 420-422*

[19] Owen and Sykes, 1987 Owen CS, Sykes NL. Magnetic labeling and cell sorting. *J Immunol Methods* 1984; *73(1):41-48.*

[20] Justin *et al*, 2017 Justin C, Philip SA, Samrot AV. Synthesis and characterization of superparamagnetic iron oxide nanoparticles (SPIONs) and utilization of SPIONs in X-ray imaging. *Appl Nanosci 2017; 7:463–475, DOI 10.1007/s13204-017-0583-x.*

[21] Justin C, Samrot AV, P. DS, Sahithya CS, Bhavya KS. Preparation, characterization and utilization of coreshell super paramagnetic iron oxide nanoparticles for curcumin delivery. *PLOS ONE 2018; 13(7): e0200440. doi.org/10.1371/journal.pone.0200440*

IOP Conf. Series: Materials Science and Engineering 1233 (2022) 012010 doi:10.1088/1757-899X/1233/1/012010

Al^{3+} and Cr^{3+} co-substituted NiZnCo nano ferrites: synthesis and structural properties

G. Vishnu Priya[1], N. Murali[2, *], J. Madhuri Sailaja[3], Vemuri Ragavendra[4], D. Parajuli[5], and P. V. Lakshmi Narayana[1]

[1]Department of Nuclear Physics, Andhra University, Visakhapatnam – 530003, A. P., India
[2]Department of Engineering Physics, AUCE (A), Andhra University, Visakhapatnam, A. P
[3]Department of Engineering Physics, Andhra University College Of Engineering For Women, Visakhapatnam, A. P., India
[4]Department of Physics, Aditya College of Engineering and Technology, Surampalem
[5]Research Center for Applied Science and Technology, Tribhuvan University, Kirtipur, Nepal

*Corresponding author email: muraliphdau@gmail.com

Abstract. The Al^{3+} and Cr^{3+} substituted in $Ni_{0.4}Zn_{0.35}Co_{0.25}Fe_{2-x-y}Al_xCr_yO_4$ (x = y = 0.0, 0.05, 0.1, 0.15 and 0.2) ferrite nanoparticles were prepared with sol-gel method. X-ray diffraction (XRD), Field Effect Scanning Microscopy (FESEM), and Fourier Transform Inferred Spectroscopy (FT-IR) were used for their characterization. They have a single-phase spinel structure. The lattice constant decreases with increasing concentrations. Agglomerated small particles with stoichiometric proportions were obtained from FESEM. The region around 600 cm^{-1} and 400 cm^{-1} has two absorption bands related to two interstitial sites by FTIR agree with the spinel lattice.

Keywords: NiZnCo nano ferrites; XRD; FESEM; FTIR.

1. Introduction

Ferrites are materials of great importance in technological applications because of their combined electromagnetic characteristics [1, 2]. Higher resistivity, magnetic saturation, permeability with lower dielectric loss etc., make them highly applicable in daily life [3]. Nickel ferrite ($NiFe_2O_4$), like ferrite, is used in soft magnets and low materials at high frequencies. $NiFe_2O_4$ has an inverse spinel structure in which Ni2+ and Fe3+ ions, respectively, occupy the octahedral B and tetrahedral A sites. There are 24 metal ions in a unit cell and 32 oxygen ions. The ionic radii of the ions determine their occupancy in the two sites [4, 5].

The properties of the ferrite are affected by the preparation method, structure of the ferrite, dopant, etc. Nickel and their substituted nickel ferrites are highly sensitive and intensively studied [6]. They are highly resistive with low losses, appropriate for microwave devices. They are environmentally stable and are highly efficient in the RF region. Some studies show that the Cr3+ ions prefer an octahedral site after substituting Fe^{3+}[7].

Similarly, Ni-Zn ferrite with Al^{3+} has been studied and its resulting properties [8, 9]. But the present researchers thought that Al^{3+} and Cr^{3+} substitution is not fully exploited and thus assumed doping the said cation at Fe^{3+}-site. Therefore, structural properties were studied for Al^{3+} and Cr^{3+} substituted in NiZnCo ferrite in this present report.

2. Experimental Details

The starting materials were analytical graded pure zinc nitrate hexahydrate, nickel nitrate, cadmium nitrate nonahydrate, copper nitrate, ferric nitrate & citric acid monohydrate with citric acid as a chelating agent for the processing of metal ions. First, the required structure is obtained at a relatively low temperature, and the solution is then made neutral with the addition of ammonia. Next, metal nitrate was mixed with distilled water and citric acid in a 1:1 molar ratio. The gel was obtained after heating continuously on the magnetic stirrer at around 150°C and dried in the atmosphere. The resultant was then sintered at 800°C for 4 hours and ground into fine particles. Finally, pellets were formed with the help of 5 tons of hydraulic pressure.

The XRD (Rigaku Miniflex II) was used for the structural properties, FESEMincorporated with CuK radiation (wavelength = 1.5406 Å) was used to analyse structural properties. Field-effect scanning electron microscopy (FESEM) images were obtained using a TESCAN, MIRA II LMH with attached EDX, Inca Oxford, for morphological and compositional properties. FTIR of Prestige was used for the functional group, respectively.

3. Results and discussion

3.1 X-ray Diffraction (XRD) studies

Phase formation and structural parameters of the synthesized materials $Ni_{0.4}Zn_{0.35}Co_{0.25}Fe_{2-x-y}Al_xCr_yO_4$ (x = y = 0.0, 0.05, 0.1, 0.15 and 0.2) ferrite nanoparticles were studied from powder XRD in the step of 2θ from 10 – 70°. As shown in Figure 1, from XRD pattern of all the synthesized samples' clear and smooth intensity peaks indicate the samples' pure phase formation and crystalline nature. All the samples crystallize into an FCC spinel structure with Fd-3m group [10]. The crystallite sizes were measured using the highest peak or the diffraction from the plane (311) by employing Scherer's formula [11]:

$$t = \frac{0.89\lambda}{\beta \cos\theta}$$

't' is the crystallite size in (nm), β is the full width at half the maximum of the highest peak (3 1 1), λ is the X-ray wavelength, and θ is the diffraction angle. Since the crystallite sizes were obtained at nanoscale levels, the synthesis method leads the nanoparticles of the synthesized samples [12]. However, the crystallite sizes are relatively smaller in response to lower synthesis temperature. Thus, the calculated nanoparticles are on average crystallite size in the range of 32.15 – 22.68 nm, whereas the lattice constants are 8.343 – 8.389 Å.

The crystallite size and the lattice parameters determine magnetic and electrical properties more or less. Thus, as shown in Table 1 and Figure 2, the lattice parameter decreases with increasing concentration, which almost obeys Vegard's law [13], and the crystallite size decreases at first, then rises, and at last decreases again. Therefore, the decrease in lattice parameter with the Al^{3+}-Cr^{3+} could be reasoned out as the replacement of higher ionic radius Fe^{3+} (0.67 Å) with lower ionic radius Al^{3+} (0.53 Å), and Cr^{3+} (0.62 Å), resulting in more volume expansion [14]. The intensities of the reflections from the dominant planes (peaks) are also presented for each of the samples synthesized and easily compared.

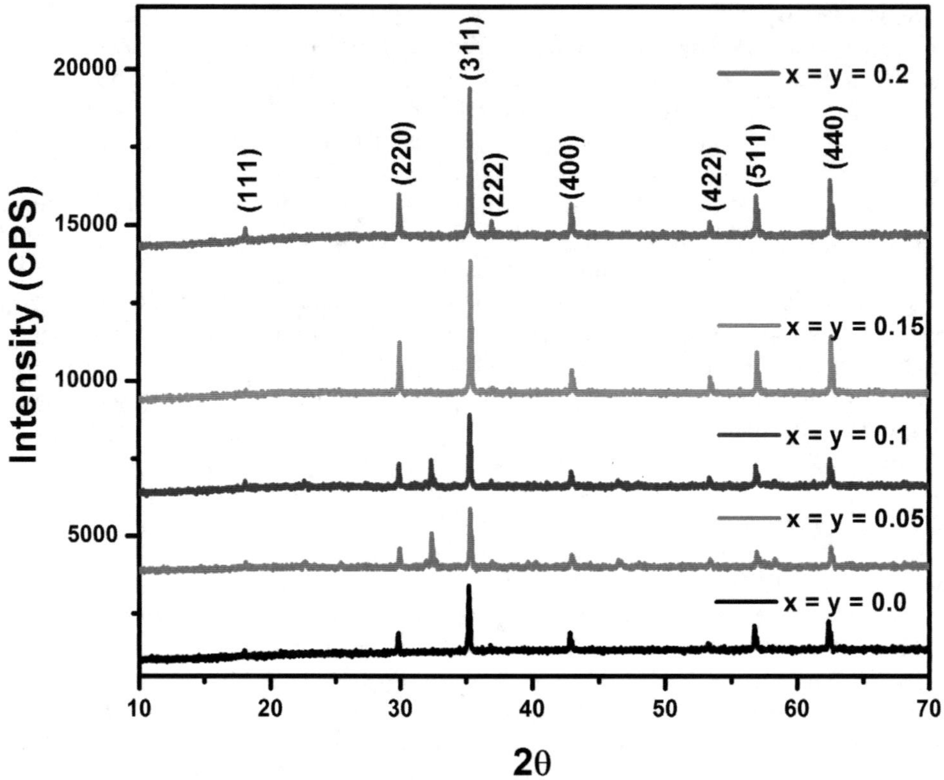

Figure 1: XRD pattern for $Ni_{0.4}Zn_{0.35}Co_{0.25}Fe_{2-x-y}Al_xCr_yO_4$ (x = y = 0.0, 0.05, 0.1, 0.15 and 0.2) ferrite nanoparticles

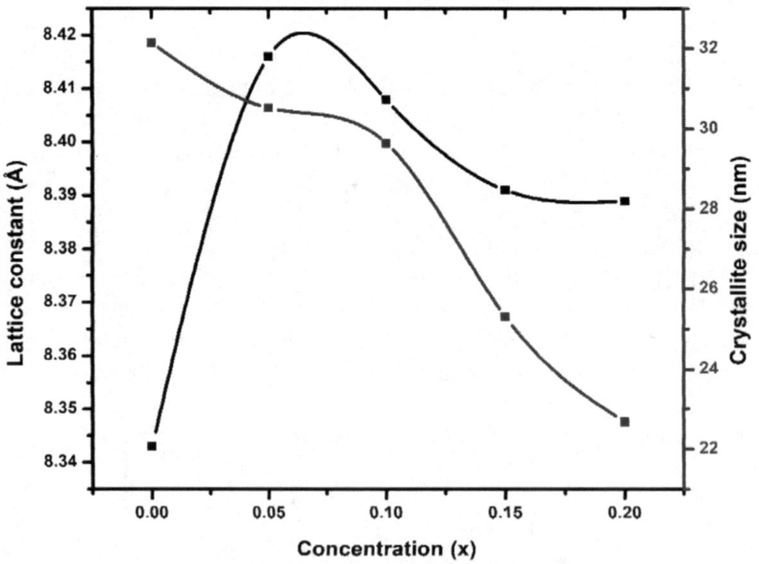

Figure 2: Lattice constant vs. crystallite size values of $Ni_{0.4}Zn_{0.35}Co_{0.25}Fe_{2-x-y}A$(x = y = 0.0, 0.05, 0.1, 0.15 and 0.2) ferrite NPs

IOP Conf. Series: Materials Science and Engineering 1233 (2022) 012010 doi:10.1088/1757-899X/1233/1/012010

Table 1: Calculated values of $Ni_{0.4}Zn_{0.35}Co_{0.25}Fe_{2-x-y}Al_xCr_yO_4$ (x = y = 0.0, 0.05, 0.1, 0.15 and 0.2) ferrite NPs.

Compound	Lattice constant a (Å)	Cell Volume (Å)^3	The crystallite size (nm)	Space Group
x = y = 0.0	8.343	611.4938	32.15	Fd-3m
x = y = 0.05	8.416	592.0142	30.52	Fd-3m
x = y = 0.1	8.408	591.6210	29.64	Fd-3m
x = y = 0.15	8.391	591.1472	25.31	Fd-3m
x = y = 0.2	8.389	590.4562	22.68	Fd-3m

3.2 FESEM Study:

The surface morphology of $Ni_{0.4}Zn_{0.35}Co_{0.25}Fe_{2-x-y}Al_xCr_yO_4$ (x = y = 0.0, 0.05, 0.1, 0.15 and 0.2) ferrite nanoparticles were studied using the FESEM. Figure 3 shows the nanograins sizes ranging from 25–64 nm and a decreasing pattern with increasing concentration. The magnetic interaction among the constituent particles causes agglomeration [15]. The morphology of the particles is very similar to compounds. They indicate that the particle size of the compounds lies in the nano-meter range having a spherical shape and a narrow size distribution. In both compounds, one can see that the grains are well separated. Also, intra-granular pores are not present on the surface of the grains. The shapes of the grains will have an impact on the physical stuff. As observed from the images, irregularly shaped grains with random orientations can be seen.

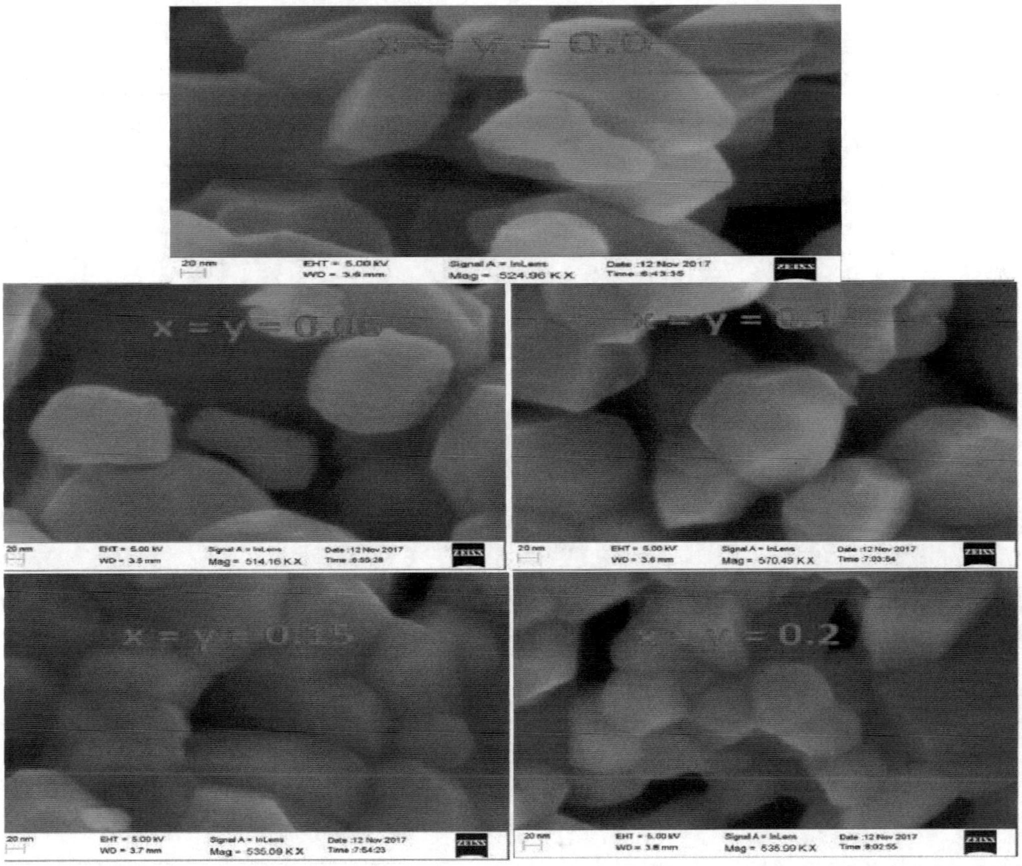

Figure 3: FESEM images for $Ni_{0.4}Zn_{0.35}Co_{0.25}Fe_{2-x-y}Al_xCr_yO_4$ (x = y = 0.0, 0.05, 0.1, 0.15 and 0.2) ferrite nanoparticles

3.3 FTIR Analysis:

The bonding nature and the cation site occupation have been studied by employing room temperature FT-IR measurements for $Ni_{0.4}Zn_{0.35}Co_{0.25}Fe_{2-x-y}Al_xCr_yO_4$ (x = y = 0.0, 0.05, 0.1, 0.15 and 0.2) ferrite nanoparticles synthesized samples. As shown in Figure 4, the two major absorption bands confirm the spinel structure of the samples [16]. The lower frequency (υ1) band around 415 – 430 cm^{-1} corresponds to a tetrahedral metal-oxygen bond, whereas the higher frequency band (υ2) around 580 – 600 cm^{-1} represents the octahedral site metal-oxygen stretching and vibrations as listed in Table 2. Though there are other absorption peaks, the dominants are the mentioned ones showing the ferrites' spinel structures. Moreover, in our case, the tetrahedral and octahedral site vibration could be inferred to the samples' constituent cations, i.e., Zn-O-Al-Cr and Ni–O vibrations, respectively [17].

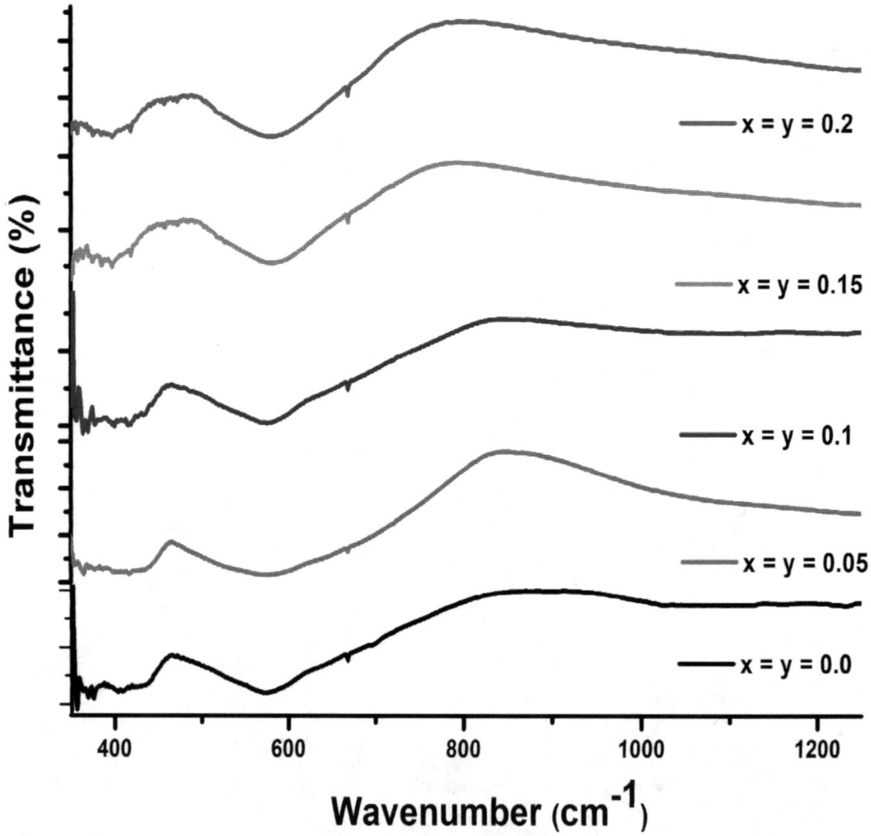

Figure 4: Infrared spectra of $Ni_{0.4}Zn_{0.35}Co_{0.25}Fe_{2-x-y}Al_xCr_yO_4$ (x = y = 0.0, 0.05, 0.1, 0.15 and 0.2) ferrite nanoparticles

Table 2: Absorption bands of $Ni_{0.4}Zn_{0.35}Co_{0.25}Fe_{2-x-y}Al_xCr_yO_4$ (x = y = 0.0, 0.05, 0.1, 0.15 and 0.2) ferrite nanoparticles

Compounds	Tetrahedral bands υ2(cm^{-1})	Octahedral bands υ1(cm^{-1})
x = y = 0.0	402.13	580.48
x = y = 0.05	405.25	581.24
x = y = 0.1	406.85	583.25
x = y = 0.15	407.12	584.25
x = y = 0.2	408.25	585.74

IOP Conf. Series: Materials Science and Engineering 1233 (2022) 012010 doi:10.1088/1757-899X/1233/1/012010

4. Conclusion

Samples with $Ni_{0.4}Zn_{0.35}Co_{0.25}Fe_{2-x-y}Al_xCr_yO_4$ (x = y = 0.0, 0.05, 0.1, 0.15 and 0.2) ferrite nanoparticles compositions have been prepared with sol-gel auto combustion method. Single-phase spinel structured ferrites with decreasing lattice parameters with Al^{3+} and Cr^{3+} substitution in Fe^{3+} content was obtained. The synthesised samples also observed randomly oriented grains with non-uniform shapes and decreasing grain sizes.

References

[1] P. Himakar, N. Murali, D. Parajuli, V. Veeraiah, K. Samatha, Tulu Wegayehu Mammo, Khalid Mujasam Batoo, Muhammad Hadi, Emad H. Raslan, and Syed Farooq Adil, Journal of Electronic Materials (2021): 1-9.
[2] Varma, P. V. S. K Phanidhar, B. Suryanarayana, Vemuri Raghavendra, D. Parajuli, N. Murali, and K. Chandramouli, Solid State Technology 63, no. 5 (2020): 8820-8827.
[3] N. Murali, S. J. Margarette, G. Pavan Kumar, B. Sailaja, S. Yonatan Mulushoa, P. Himakar, B. Kishore Babu, and V. Veeraiah, Physica B: Condensed Matter 522 (2017): 1-6.
[4] S. Kuswanth Kumar, N. Murali, D. Parajuli, A. Ramakrishna, P. S. V. Subba Rao, and M. P. Rao, Solid State Technology 63 (5) (2020) 4069-4077.
[5] S. Yonatan Mulushoa, Ch. Vijaya Kumari, Vemuri Raghavendra, K. Ephraim Babu, B. S. N. Murthy, Kundavarapu Suribabu, Y. Ramakrishna, and N. Murali, Physica B: Condensed Matter 572 (2019): 139-147.
[6] Chandan Upadhyay and H. C. Verma, Journal of Applied Physics, 95 (10), (2004), 5746–5751.
[7] M. Abdullah Dar, Jyoti Shah, W. A. Siddiqui, R. K. Kotnala, Appl Nanosci (2014) 4:675–682.
[8] Monica Sorescu, L. Diamandescu, R. Peelamedu, R. Roy, P. Yadoji, Journal of Magnetism and Magnetic Materials 279 (2004) 195–201.
[9] K. Rama Krishna, K. Vijaya Kumar, C. Ravindernathgupta, Dachepalli Ravinder, Advances in Materials Physics and Chemistry, 2012, 2, 149-154.
[10] Tsoncheva, T., Manova, E., Velinov, N., Paneva, D., Popova, M., Kunev, B., Tenchev, K., Mitov, I.: Catal. Commun. 12, 105–109 (2010).
[11] Lasheras, X., Insausti, M., Gil De Muro, I., Garaio, E., Plazaola, F., Moros, M., De Matteis, L., De La Fuente, J.M., Lezama, L.: J. Phys. Chem. C. 120, 3492–3500 (2016).
[12] Poh Lin Leng, Mahmoud Goodarz Naseri, Elias Saion, Abdul Halim Shaari, Mazaliana Ahmad Kamaruddin, Advances in Nanoparticles, 2013, 2, 378 – 383.
[13] M. R. Patil, M. K. Rendale, S. N. Mathad, and R. B. Pujar, Inorganic and Nano-Metal Chemistry, 47 (8), (2017), 1145–1149.
[14] Rajinder Kumar, Ragini Raj Singh, and P. B. Barman, International Journal of Scientific & Engineering Research, 5 (5), (2014), 12–20.
[15] D R S G Swamy, M C. Varma, K.S Rao, K.H. Rao, Indian Journal of Research in Pharmacy and Biotechnology, 2014, 59 – 67.
[16] Ovidiu F. Caltun, Leonard Spinu, and Alexandru Stancu, IEEE transactions on magnetics, 37 (4), (2001), 2353–2355.
[17] B. H. Devmunde, A. V. Raut, S. D. Birajdar, S. J. Shukla, D. R. Shengule, and K.M. Jadhav, Journal of Nanoparticles Volume 2016, 4709687.

Giant dielectric behavior in $NdCrO_3$ perovskite

Jada Shanker[1,*], J Ananthaiah[1], N Pavan Kumar[2], Kasarapu Venkataramana[3], M Anand pandarinath[3], Ujwal U P[4], P. Raju[5], E Meher Abhinav[1] and D Suresh babu[6]

[1]Department of Humanities and Sciences, K.G Reddy College of Engineering and Technology, Chilkur (V), Moinabad (M), RR Dist, Telangana, India
[2]Department of Sciences and Humanities, Matrusri Engineering College, Saidabad, Hyderabad, Telangana, India
[3]Keshav Memorial Engineering College, Hyderabad-500058, Telangana, India.
[5]Department of Science and Humanities, Vidya Jyothi Institute of Technology, Hyderabad
[4]Department of Science and Humanities, Rajarambapu Institute of Technology, Sangli, Maharastra-415414, India
[5]Freshman Engineering, Geethanjali College of Engineering and Technology, Cheeryala (V), Keesara (M), Medchal Dist., Telangana, India
[6]Department of Physics, University College of Science, Osmania University, Hyderabad

**Corresponding author:dr.jadashanker.physics@gmail.com*

Abstract. $NdCrO_3$ perovskite compound synthesized via sol-gel process. Impedance spectroscopic data have been studied in the wide frequency range (0.1Hz -10MHz), at various temperatures (RT-500°C). Structural analysis have done by X-ray diffractometer, it indicates single-phase orthorhombic structure. Particle size is found to be 32nm. Microstructure seems to be uniform distribution of grains throughout the surface. This material showed the giant dielectric behaviour ($\varepsilon' > 10^4$). Impedance studies of $NdCrO_3$ noticed that it is negative temperature coefficient of resistance type semiconductor.

Keywords: Sol-gel auto combustion technique, Perovskite nanoparticles, Impedance, giant dielectric, and ac conductiv ity.

1. Introduction:

ABO_3 perovskite polycrystalline compounds show well-known properties like colossal magneto-resistance, high-temperature superconductivity (Tc) and multi-ferocity. etc. **[1-2]**. $RCrO_3$ Orthochromaties have considering for devices applications like solid-oxide fuel cells and catalytic converters. SOFC properties like high electrical conductivity, high densification, and good chemical stability were found in

$NdCrO_3$ material **[2].** In recent years, $RCrO_3$ type of materials exhibited high dielectric constant ($\varepsilon' > 10^4$), a so-called giant dielectric constant (ε') **[3]**.

2. Experimental procedure:

$NdCrO_3$ Polycrystalline material was prepared by sol-gel auto combustion technique and has been reported **[4].** The obtained powder samples were annealed at 750^0C for 4 h and then pelletized with uniaxial pressing. In the polycrystalline ceramics, sintering route is essential that effects on microstructure, grain growth and densification **[5].** So that, these cermic pellets have been sintered in air atmosphere at 450^0C temperature for 4 h. The SEM and dielectric measurements were done for these samples.

3. Results and Discussion:

3.1. X-ray diffraction analysis

XRD Rietveld refinement patterns of $NdCrO_3$ sample sintered at 450^oC has been analyzed using fullproof software, it is noticed that single-phase orthorhombic structure with Pnma space group. Lattice parameters, refinement parameters have evaluated and are reported **[4]**.

3.2. Scannining electron mictroscopy analysis

Figure 1 shows the grain size, grain spreading and voids in microstructure of $NdCrO_3$. The grains are in homogeneously dispersed around the surface, and more gaps are presenting in the material, it indicating that pellet has some quantity of porous may be due to the low sintering temperature. From the SEM images, it is noticed that grains have agglomerated due to the porous in material bounded with grains **[5].** The SEM microstructure also shows the changing grain sizes and well-defined grain boundaries reveal the $NdCrO_3$ is polycrystalline behaviour **[6].** Density of the $NdCrO_3$ polycrystalline pellets have been estimated and found to be 84%.

Figure 1. Scanning electron microscopy (SEM) Microstructure of sintered pellet of $NdCrO_3$ sintered pellet.

3.3. Dielectric analysis

Dielectric constant (ε') as function of frequency of $NdCrO_3$ at various temperatures has shown in **Figure 2**. Dielectric constant (ε') decreased as increase in frequency for all specified temperatures. High dielectric constant has seen that in the low-frequency region because of the interfacial polarizations, dislocations, oxygen vacancies, pile ups, grain boundary influence, etc **[7]**, ε' is decreasing with rising frequency and merged at high frequency region (10^7Hz) due to lowering the barrier properties subsequently, which leads to develop orientational polarization [**8-9**]. The high value of dialectic constant (ε'>10^4) in the low-frequency region so called giant dielectric nature **[3].** The dielectric constant (ε') is augmented with rising temperature (RT-500°C), and it grows into more prominent at low frequencies. It indicates the thermally stimulated nature in material [**9**]. Dielectric constant and dielectric loss have gradually decreased with rising frequency; it can be explained using polarization mechanisms **[10].** In the lower frequency region dipoles can displace by alternating electric field, rhence it develops higher value of dielectric constant, whereas at high frequency region dipoles cannot displace rapidly with changing field, hence dielectric constant value is decreasing.

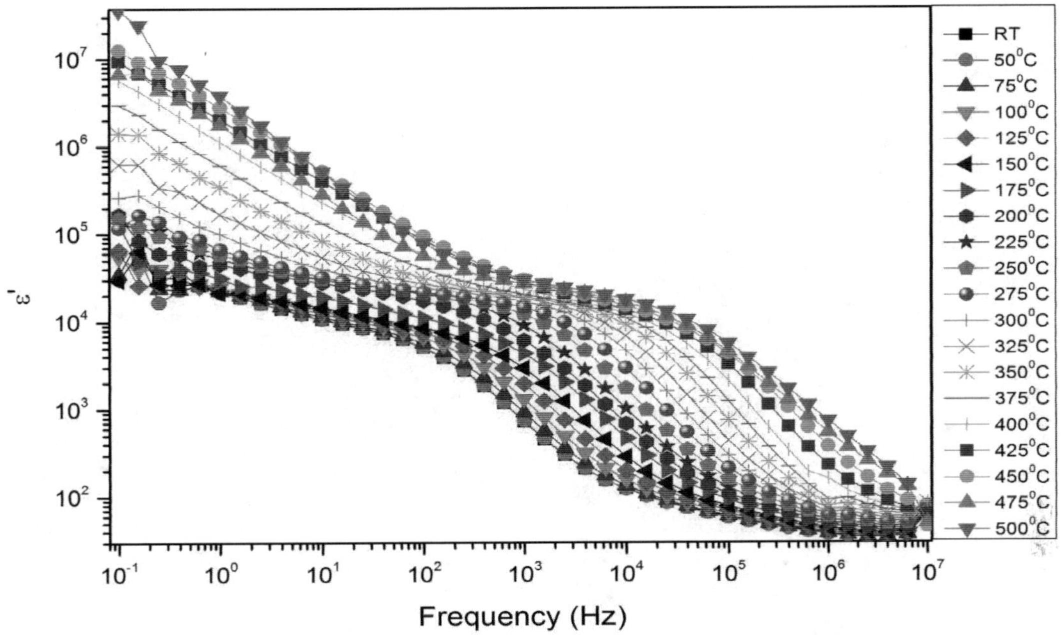

Figure 2. Frequency dependence of the dielectric constant (ε') of $NdCrO_3$ at different temperatures.

Figure 3 shows frequency dependent dielectric loss ($tan\delta$) of $NdCrO_3$ perovskite at various temperatures. Tanδ declines as grow in frequency for given temperature. As growing the temperature, tanδ upsurges in low frequencies which corresponds to space charge polarization **[11].** The dielectric loss plot shown broad peak in each curve, and are shifting towards higher frequencies with increasing temperature. The dielectric loss (tanδ) is decreasing at higher temperature, which reveals the presence of dielectric relaxation. The dielectric loss ($tan\delta$) is more significant in the low frequency region, suggesting the low conductivity of the grain boundaries. In high frequency region loss is unchanging due to high conductivity of grains **[12].**

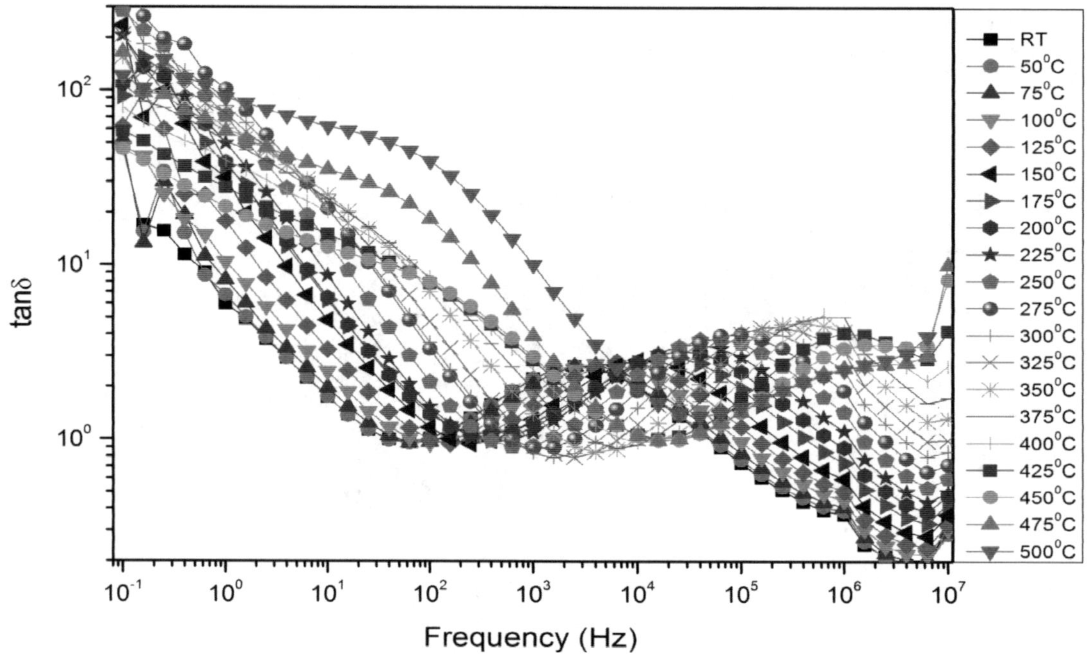

Figure 3. Frequency dependence of the dielectric loss ($tan\delta$) of $NdCrO_3$ at different temperatures.

3.4. Impedance analysis

Frequency variation real part of impedance (Z') of $NdCrO_3$ at different temperatures presented in **Figure 4**. Z' is more prominent at lower frequencies, and gradually decreased as increase in frequency and temperature, it indicates semiconductor behavior **[3]**. Z' decreased with rise frequency, which leads to rise in ac conductivity (σ_{ac}) as increase in frequency. All the Z' curves ware merged independent of temperature at high frequency region, suggesting the possible discharge of space charges and subsequently dropping the hurdle properties in material **[13-15].**

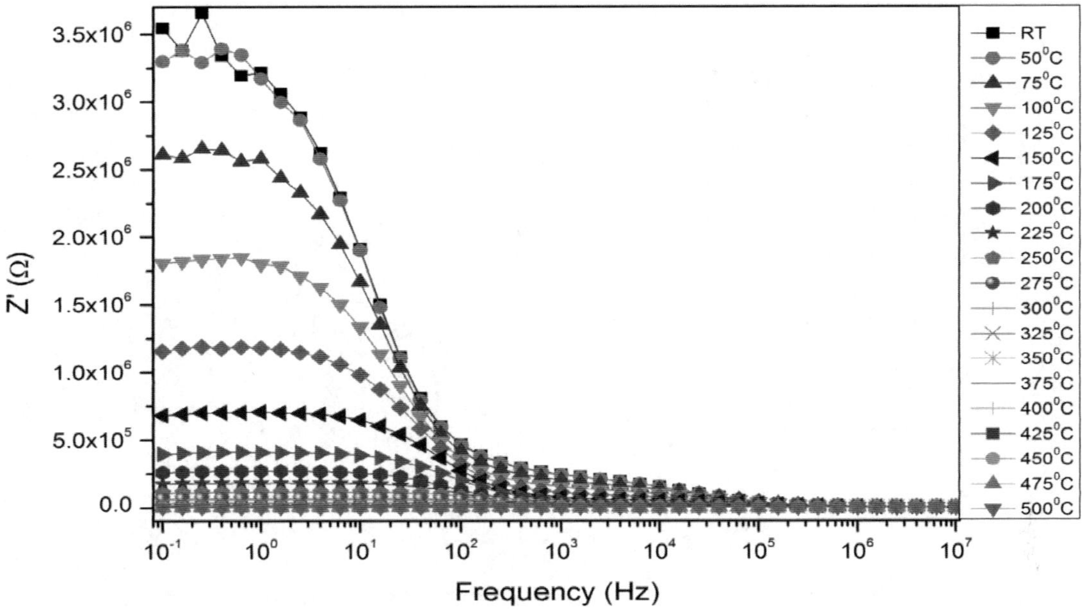

Figure 4. Frequency dependence of Real part impedance (Z') of $NdCrO_3$ at different temperatures

Figure 5 shows imaginary impedance (Z'') as function of frequency of $NdCrO_3$ perovskite at different temperatures. Two prominent peaks are seen in each curve one at a high-frequency region and other one at low-frequency region, these corresponds to grain and grain boundary relaxations respectively **[16]**. Both the peaks are shifting towards the higher frequencies as increasing the temperature due to the presence of an electrical relaxation, suggesting temperature reliant relaxation. Full width half maxima of peaks are found to decline gradually as grow in frequency reveals the increasing of conductivity of grain and grain boundary. The shifting of peak points suggesting the possibility of a distribution of relaxation time.

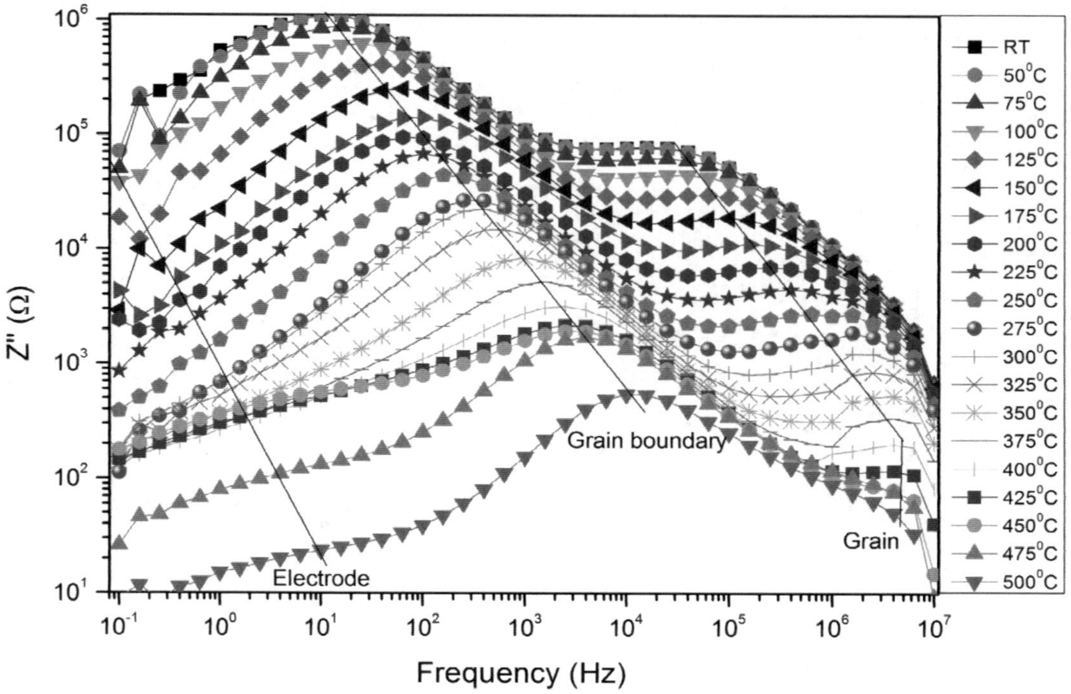

Figure 5. Frequency dependence of imaginary impedance (Z″) of $NdCrO_3$ at different temperatures.

4. Conclusions:

$NdCrO_3$ perovskite ceramic sample has been prepared by Sol-gel process. This material show single phase orthorhombic configuration with Pnma space group. Impedance investigation noticed that this material have semiconducting nature. This material shows the giant dielectric behavior ($\varepsilon' > 10^4$), and is exhibited temperature and frequency dependent relaxation. These results suggesting that $NdCrO_3$ is suitable for gaint capacitors applications.

Acknowledgements:

Authors would like to thank to Dr. M. Buchi Suresh, ARCI, Hyderabad, Telangana, India for the providing experimental facility to carry out this work.

References:

[1]. Yi Du, Zhen Xiang Cheng, Xiao-Lin Wang, and Shi Xue Dou "Structure, magnetic, and thermal properties of $Nd_{1-x}La_xCrO_3(0 \leq x \leq 1)$ *Journal of Applied Physics* **108**, 093914 (2010).

[2]. Yu Shen, Monan Liu, and Tianmin He, San Ping Jiang. "Preparation, Electrical Conductivity, and Thermal Expansion Behavior of Dense $Nd_{1-x}Ca_xCrO_3$ Solid solutions" *J. Am. Ceram. Soc.* **92** [10] 2259-2264 (2009).

[3]. Bandi Vittal Prasad, G.Narsinga Rao, J.W. Chen, D. Suresh Babu. "Relaxor Ferroelectric like giant Permittivity in $PrCrO_3$ semiconductor ceramics" *Materials Chemistry and Physics* **126** (2011) 918-921.

[4]. Jada Shanker, M.Buchi Suresh, D.Suresh babu "Synthesis, Characterization and Electrical Properties of $NdXO_3$(X=Cr, Fe) Nanoparticles" *Materials Today: Proceedings 3* (2016) 2091-2100.

[5] Leow Chun Yan, Jumian Hassan, Mansor Hashim, Wong Swee Yin, Tan Foo Khaoon and Wong Yick Jeng, "Effect of Sintering Temperatures on the Microstructure and Dielectric Properties of SrTiO*3*" *World Applied sciences Journal* **14**(7):1091-1094, 2011

[6]. H.O. Rodrigues, G.F.M. Pires Junior, J.S. Almeida, E.O. Sancho, A.C. Ferreira, M.A.S. Silva, A.S.B. Sombra."Study of the structural, dielectric and magnetic properties of Bi_2O_3 and PbO addition on $BiFeO_3$ ceramic matrix" *Journal of Physics and Chemistry of Solids* **71**(2010) 1329-1336

[7]. A. Mitra, A.S. Mahapatra, A. Mallick, A. Shaw, N. Bhakta, P.K. Chakrabarti, Improved magneto-electric properties of $LaFeO_3$ in $La_{0.8}Gd_{0.2}Fe_{0.97}Nb_{0.03}O_3$, *Ceramics International* Volume **44**, Issue 4, March 2018, Pages 4442-4449

[8]. S. Sreehari Sastry, L. Tanuj Kumar, and Ha Sie Tiong "Dielectric Studies on Benzothiazole Based Liquid Crystals at Radio Frequency Region" *International Journal of Innovative Research in Science, Engineering and Technology*, Vol. **3**, Issue 3, March 2014 ,10212-10219

[9]. Subrat K. Barik, R.N.P. Choudhary, A.K. Singh "AC impedance spectroscopy and conductivity studies of $Ba_{0.8}Sr_{0.2}TiO_3$ ceramics" *Adv. Mat. Let.* 2011,**2**(6),419-424.

[10]. Shuai Huang, Liran Shi, Zhaoming Tian, Songliu Yuan, Liguang Wang, Gaoshang Gong,Chongyang Yin, Gebru Zerihun "High-temperature colossal dielectric response in $RFeO_3$(R=La, Pr and Sm) ceramics" *Ceramics International* **41** (2015) 691-698.

[11]. Shrivastav B.D., Barde R., Mishra A. And Phadake S. "Frequency and Temperature Dependence of Dielectric Properties of Fish Scales Tissues" *Research Journal of Physical Sciences,* Vol. **1** (6), 24-29, July (2013)

[12]. Navneet Sing, Ashish Agarwal, Sujata Sanghi. "Dielectric relaxation, conductivity behavior and magnetic properties of Mg Substituted Zn-Li ferrites" *Current Applied Physics* **11** (2011) 783-789.

[13]. Hemant Singh, Amit Kumar, K.L. Yadav, "Structural, dielectric, magnetic, magnetodielectric and impedance spectroscopic studies of multiferroic $BiFeO_3$–$BaTiO_3$ ceramics" *Materials Science and Engineering B* **176** (2011) 540–547.

[14]. Archana Shukla, R.N.P. Choudhary, "Impedance and modulus spectroscopy characterization of La þ 3/Mn P4 modified $PbTiO_3$ nanoceramics" *Current Applied Physics* **11** (2011) 414e422.

[15]. Balgovind Tiwari, R.N.P. Choudhary, "Frequency-temperature response of $Pb(Zr_{0.65-x}Ce_xTi_{0.35})O_3$ ferroelectric ceramics: Impedance spectroscopic studies" *Journal of Alloys and Compounds* **493** (2010) 1–10

[16]. S. A. Acharya, K. Singh "Grain shape effect on dielectric properties and ionic Conductivity of Gd-doped ceria" *Adv. Mat. Let. 2014*, **5**(2), 61-66.

IOP Conf. Series: Materials Science and Engineering 1233 (2022) 012012 doi:10.1088/1757-899X/1233/1/012012

Microstructural and mechanical characterization of HSLA X70 Welded joints using eddy current testing (ECT)

N. Bensaid[1], M. F. Benlamnouar[1], R. Badji[1], T. Saadi[1], M. Zidani[2], M. Gousmine[1].

[1] Research Center in Industrial Technologies CRTI cheraga 16014, Algiers, Algeria.
[2] Laboratoire de Génie Énergétique et Matériaux, BP 145, Biskra 07000 Algérie.

Abstract. This paper proposes an approach for measuring the mechanical properties by non-destructive testing techniques using eddy current testing (ECT), for a welded and heat-treated microalloyed material (X70) dedicated to hydrocarbon transport. This study offers a new measurement method based on linear approach, to expect mechanical property in microstructures by using Non-destructive testing (ECT) that can apply to mechanical systems, materials, weldments, etc. The result show that hardness measurements are in broad have the same trend with impedance function which depend of heat treatment variable applied to weld joint.

1. Introduction

Oil transport under production conditions requires the use of very high pressures, and therefore the use of high strength HSLA steels with good welds [1-2].

Welding procedure is the principal metal joining process and is widely used for oil industry. It is an interdisciplinary field mobilizing electric, thermal and thermo-mechanical phenomena at the same time [3, 5]. In addition, the safety of welded installations requires good mechanical property to prevent bursting and damage to pipelines.

Indeed, it during the welding operation, the residual stresses state and microstructural evolution of weld joint cause a reduction in weld characteristics, mainly on the mechanical properties [6, 7]. Thus, the need to characterize weld damage is real, through easily methods in a short time, which permit it will not affect or cut the welded pipe structure to take specimens.

In common, to ensure the reliability of a pipe, it is necessary to rely on different techniques such as NDT and numerical methods [8, 9]. Non-destructive testing is an overview of the different techniques to meet the needs of high-tech sectors such as the pipeline industry. One of these NDT techniques is eddy current [9].

The ECT of materials is significantly related to their chemical composition, their metallic phases and their residual stresses caused by thermal or thermo mechanical processes [5]. Consequently, it is therefore important to use ECT tests to establish relationship with one of the mechanical characteristics of weld joint to evaluate weld damage. In this context, this study investigates microstructural and mechanical behavior of HSLA X70 Welded joints using eddy current testing (ECT), to perform the results, Different specimens states (fusion zones, heat affected zone and base metal), were characterized around weld joints using post weld heat treatment. Results were compared with those obtained by ECT and hardness test.

2. Experimental procedure

In this work, the HSLA-X70 base material is taken from pipeline of 711.2mm in diameter and 11.48 mm thick (Table 1.). The chemical compositions of base and filler metal, measured by FOUNDRY-MASTER spectrometer, are respectively given in Table 1 and 2. Welding process was carried out in perpendicular direction to rolling direction of BM with square butt joints. Welding process has been performed using shielded metal arc welding in 05 passes based on ASME standard (Table 3) [10].

Table.1. Chemical composition of base metal.

	C	Mo	Cu	Ni	Mn
Element %	0.075	0.016	0.162	0.031	1.79
	Cr	**P**	**Si**	**V**	**S**
	0.055	0.016	0.335	0.010	0.0048

Table.2. Chemical composition of filler material **(E9010-G)**.

Element	C	Si	Mn	Mo	Ni	Cr	Cu
W %	0.07	0.4	1.5	0.35	0.15	0.15	0.05

Table.3. Welding process parameters.

Number of passes	1	2	3	4	5
Preheating (° C)	170	195	200	197	175
Arc voltage U (V)	27	25	24	26	23
Welding current I (A)	83	138	144	161	143
Welding speed (mm/min)	42	96	71	71	120

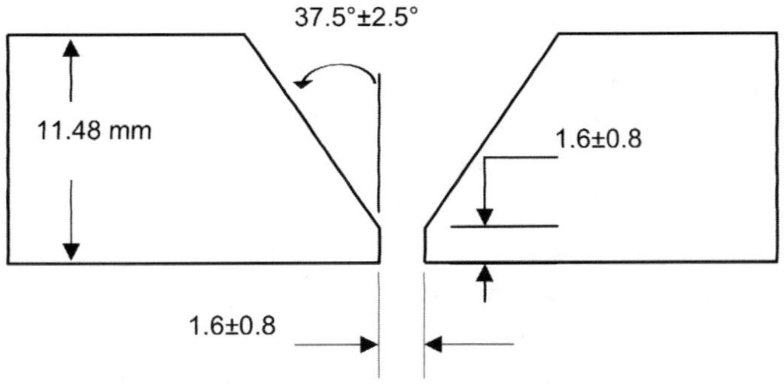

Figure.1 Welding sketch

IOP Conf. Series: Materials Science and Engineering 1233 (2022) 012012 doi:10.1088/1757-899X/1233/1/012012

2.1. The samples used

We considered rectangular samples 3mm thick and 10mm deep and 11mm high of HSLA X70, taken perpendicular to the direction of welding, such that the distance between each sample is 5mm, these are drawn for each weld zone; these samples were subdivided into 6 types (figure 2).

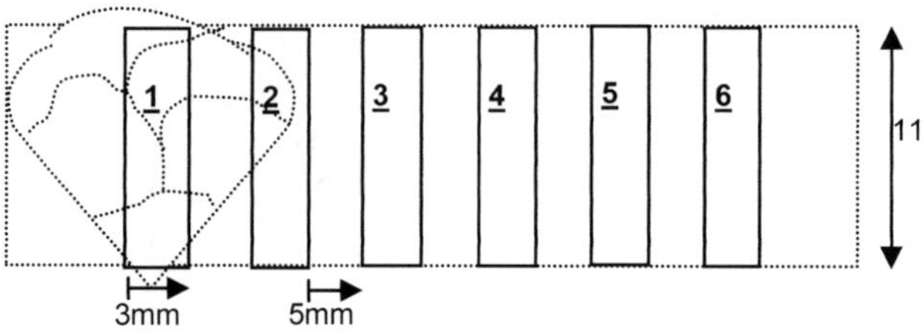

Figure.2 Taking specimen: 1 (ZF), 2 (ZF+ZAT), 3 (ZAT), 4 (MB1), 5 (MB2) 6 (MB3)

2.2. Polishing procedures
Standard polishing procedure was used for optical metallography. The specimens were then etched with a Nital solution with 3% of concentration for 3min. The microstructural examination of the samples was characterized by ZEISS optical microscope.

2.3. Impedance
By using bridge methods, the impedance is measured of the current and magnitude of voltage, and the difference values from them, calibrated reference impedance is set to reveal the effect of the impedance sample under testing.

2.4. Heat treatment
A post weld heat treatment was carried out by using annealing at 650 °C for 30 minutes with slow re-cooling to minimize residual stresses generated by welding thermal cycles [11-13].

2.5. Vickers Hardness
Vickers Hardness tests were conducted by HWDM-1 instrument under a load of 10Kgf and 1 min acquisition time. Each Hardness measurement was performed by 05 try to ascertain hardness value at the middle of the weld joint, perpendicularly to the welding direction.

3. Result and discussion

3.1. Microstructural characterization
The figure 3 shows the macrography of weld joint. The microstructures of weld joints (BM, HAZ, FZ) are presented in (Figure 4, 5 and 6) respectively. The microstructure of base metal is composed of polygonal ferrite α (white color) and pearlite P (black color), containing about 80% ferrite with a range of ferrite grain sizes between 4 to 8µm (Figure 4). The microstructure of the (ZAT) undergoes a decrease in grain size when it becomes close to the base metal, the latter is distinguished by a softer pearlitic-ferritc structure with higher ferrite content (Figure 5).
Fusion zone was partially quenched and composed mainly of martensite (ά), acicular ferrite(α), a small amount of benite (B), residual austenite (γ) and dispersed carbide particles (Figure 6).
In general PWHT treatment conducted resulted in a slight ferrite grain growth in BM, FZ and HAZ (Show Figures 4, 5, 6).

IOP Conf. Series: Materials Science and Engineering 1233 (2022) 012012 doi:10.1088/1757-899X/1233/1/012012

Figure.3. Macrographique of weld joint

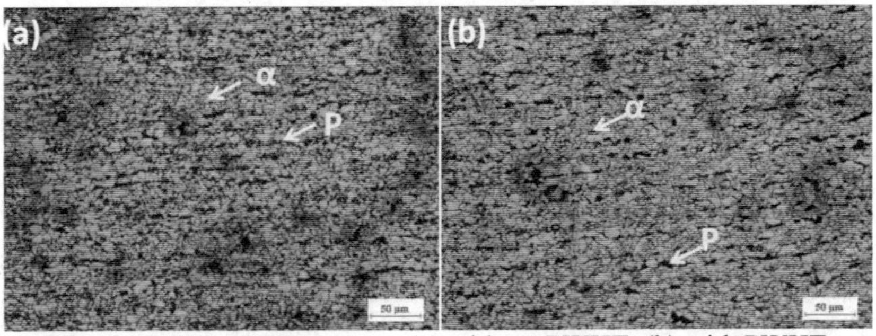

Figure.4. BM microstructure, (a) without PWHT, (b) with PWHT

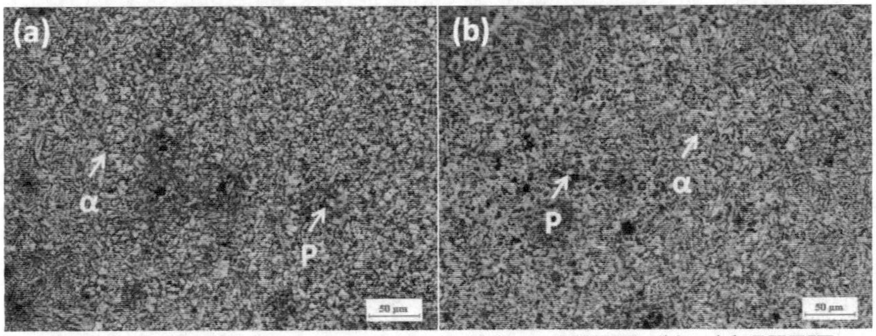

Figure.5. HAZ microstructure, (a) without PWHT, (b) with PWHT

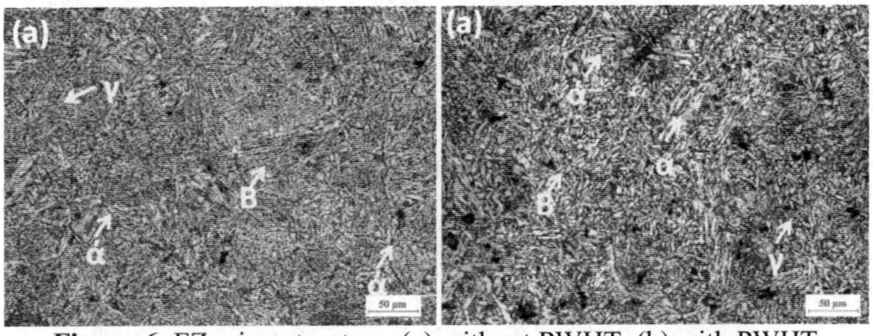

Figure.6. FZ microstructure, (a) without PWHT, (b) with PWHT

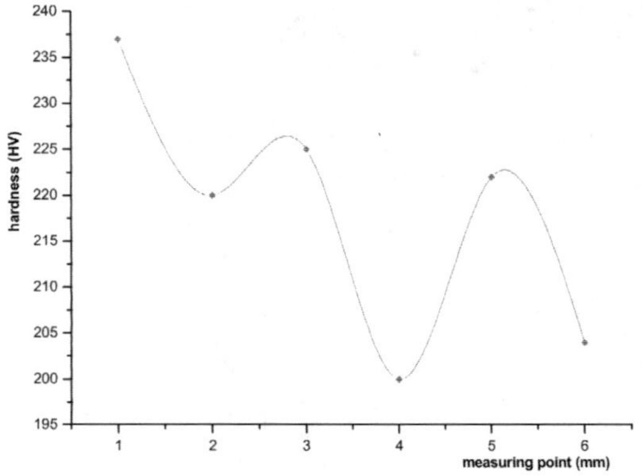

Hardness (HV) of weld joint without PWHT

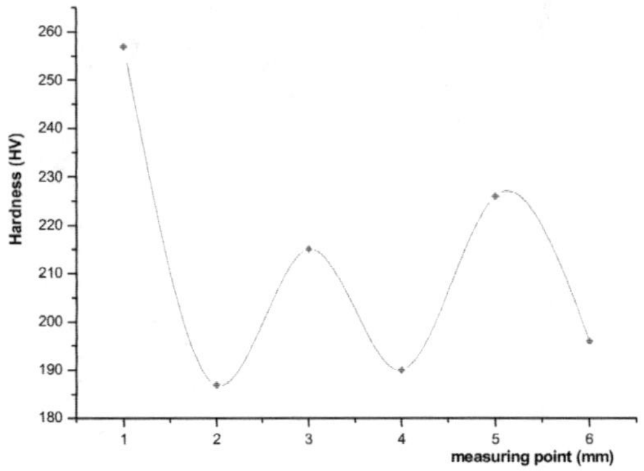

Hardness (HV) of weld joint with PWHT

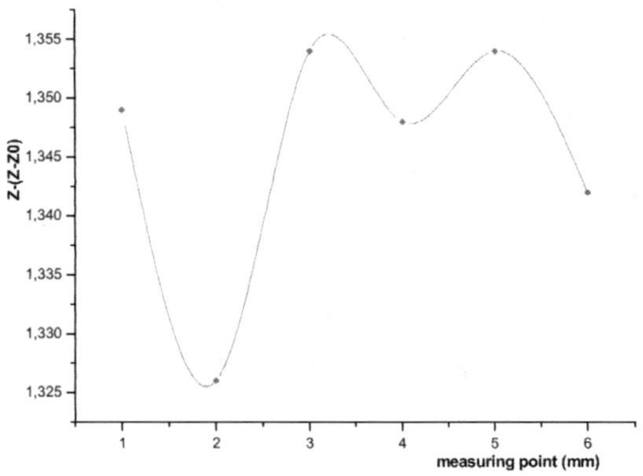

Impedance of weld joint without PWHT

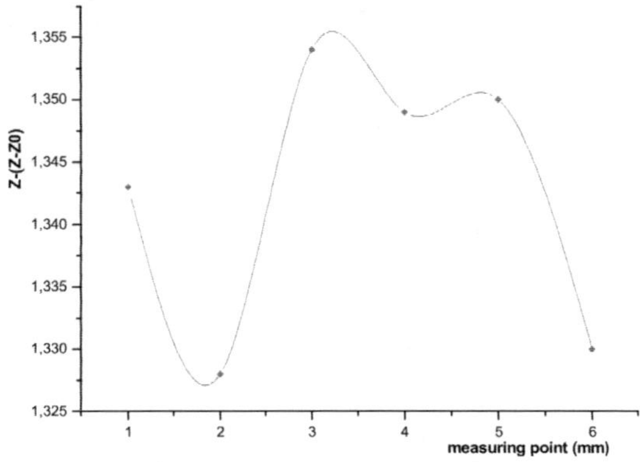

Impedance of weld joint with PWHT

Figure.7 Comparison between impedance and microhardness measurements

From Figure 7, it was noted that the heat treatment increased the microhardness of the welded joint. The rate of homogenization of the microdures and the intervals of the impedance in the untreated case are lower than those obtained in the treated case. The change in impedance follows the same pattern as the microhardness in both cases.

4. Conclusion

This work is part of a research approach to integrate a mechanical characteristic (microhardness) into NDT analysis systems.

In this study the effect of PWHT treatment on hardness and magnetic impedance of HSLA X70 weld joint was investigated. As a result, the PWHT treatment conducted resulted in a slight ferrite grain growth in BM, FZ and HAZ.

The obtained results make it possible to exercise ECT control over weld joints and pipelines to characterize mechanical behavior and weld damage in each point of the welded assembly.

A good correlation was obtained for the two presented methods (mechanical and electromagnetic tests). It concluded that, the impedance corresponds to the reorganisation of hardness of HSLA X70 weld joint under the relaxation of residual stress due to the effect of post weld heat treatment

Ultimately, the methods developed can make it possible not only to determined microhardness trough eddy current analysis, but also to create new programs and devices to identificate of mechanical properties by electromagnetic measurements.

IOP Conf. Series: Materials Science and Engineering 1233 (2022) 012012 doi:10.1088/1757-899X/1233/1/012012

Références

[1] M.F. Benlamnouar, M. Hadji, R. Badji, N. Bensaid, T. Saadi, Y. Laib dit Laksir, S. Senouci. Optimization of TIG Welding Process Parameters for X70-304L Dissimilar Joint Using Taguchi Method. Solid State Phenomena, vol. 297, pp. 51-61, 2019.

[2] S.H. Hashemi, D. Mohammadyani, M. Pouranvari, S.M. Mousavizadeh, On the relation of microstructure and impact toughness characteristics of DSAW steel of grade API X70, Fatigue Fract. Eng. Mater. Struct., vol. 32, pp. 33–40, 2009.

[3] L. Zhongqiu, F. Jian, Z. Yong, Y. Zexi, Influence of quenching on-line on properties of X70 steel for sour service seamless pipe. Energy Proc., vol. 16, pp. 444–450, 2012 .

[4] Anshul Yadav, Aniruddha Ghosh, Arvind Kumar, Experimental and numerical study of thermal field and weld bead characteristics in submerged arc welded plate, Journal of Materials Processing Technology, Vol. 248 (2017), pp.262-274.

[5] H. Alipooramirabad, A. Paradowska, R. Ghomashchi, M. Reid. Investigating the effects of welding process on residual stresses, microstructure and mechanical properties in HSLA steel welds. Journal of Manufacturing Processes, vol. 28,pp 70–81, 2017, doi: 10.1016/j.jmapro.2017.04.030.

[6] N. Bensaid, M. Hadji, R. Badji, M.F. Benlamnouar, T. Saadi, Y. Laib dit Laksir, S. Senouci: Solid State Phenomena, Vol. 297, 2019, p. 195-203.

[7] T. Saadi, M F. Benlamnouar, N. Bensaid, A. Boutaghane, M A. Soualili, H. Hachemi: Defect and Diffusion Forum .2021Volume 406. pp. 319-333.

[8] M F. Benlamnouar, T. Saadi, N. Bensaid, M. Gousmine, Y. Touggui, M. Temmar : Lecture Notes in Mechanical Engineering-Springer, 2021, pp. 463-472.

[9] M Zergoug, S Lebaili, H Boudjellal, A Benchaala, Relation between mechanical microhardness and impedance variations in eddy current testing, NDT & E International, Vol. 37-1(2004), pp. 65-72

[10] ASME “Qualification standard for welding and brazing” Boiler and Pressure Vessel Committee on Welding and Brazing”, the American Society of Mechanical Engineers, july 1, 2010.

[11] ASTM, ‘‘Specification for Carbon Structural Steel’’,American Society for Testing and Materials; 2005.

[12] ASM International. Fundamentals of the Heat Treating of Steel.Practical Heat Treating, Second Edition 2006.

[13] Jae-Myeong Jang, Sung-Joon Kim, Nam Hyun Kang, Kyung-Mox Ch, and Dong-Woo Suh1. Effects of Annealing Conditions on Microstructure and Mechanical Properties of Low Carbon, Manganese Transformation-Induced Plasticity Steel. Met. Mater. Int., vol. 15, no. 6, pp. 909-916, 2009

AUTHOR INDEX